Start-Up
A Technician's Guide
Third Edition

Start-Up
A Technician's Guide

Third Edition

Diane R. Barkin

ISA Technician Series

Notice

The information presented in this publication is for the general education of the reader. Because neither the author nor the publisher has any control over the use of the information by the reader, both the author and the publisher disclaim any and all liability of any kind arising out of such use. The reader is expected to exercise sound professional judgment in using any of the information presented in a particular application.

Additionally, neither the author nor the publisher has investigated or considered the effect of any patents on the ability of the reader to use any of the information in a particular application. The reader is responsible for reviewing any possible patents that may affect any particular use of the information presented.

Any references to commercial products in the work are cited as examples only. Neither the author nor the publisher endorses any referenced commercial product. Any trademarks or tradenames referenced belong to the respective owner of the mark or name. Neither the author nor the publisher makes any representation regarding the availability of any referenced commercial product at any time. The manufacturer's instructions on the use of any commercial product must be followed at all times, even if in conflict with the information in this publication.

Copyright © 2020 International Society of Automation (ISA)
All rights reserved.

Printed in the United States of America.
10 9 8 7 6 5 4 3 2

Print ISBN: 978-1-945541-79-7
Kindle ISBN: 978-1-945541-80-3
ePub ISBN: 978-1-945541-81-0

No part of this work may be reproduced, stored in a retrieval system, or transmitted in any form or by any means, electronic, mechanical, photocopying, recording or otherwise, without the prior written permission of the publisher.

ISA
67 T. W. Alexander Drive
P.O. Box 12277
Research Triangle Park, NC 27709

Library of Congress Cataloging-in-Publication Data in process

*This book is dedicated to my parents,
Jindra Brumlik, and Stanley Barkin, both of whom
instilled in me a sense of purpose,
a need for continued learning, and a strong work ethic.
They will forever be my influence and heroes.*

Contents

About the Author xiii

Chapter 1 The Role of the Control Systems Technician 1
    Project Execution and Control (Project Management) Overview 3
    1.1   Job Descriptions 4
    1.2   CST Roles 6
          1.2.1 Training 7
          1.2.2 Safety Inspector 7
          1.2.3 Liaison 8
          1.2.4 Technician Duties 9
          1.2.5 Inspection (QA/QC) 10
                  Housekeeping 10
                  Wiring Runs and Terminations 11
                  Wiring Tags 11
                  Instrumentation 12
                  PASs 12
                  Documentation 13
          1.2.6 Design and Engineering Assistance 13
          1.2.7 Leader 13
    1.3   CST Tasks 14
          1.3.1 Baseline Duties 14
    Summary 15
    Review 15
    Recommended Reading 16

Chapter 2 Applicable Safety Practices and Standards **17**
    2.1 The 14 Points of OSHA PSM. 20
        2.1.1 Point 1: Employee Involvement 21
        2.1.2 Point 2: Process Safety Information. 21
        2.1.3 Point 3: PHA 22
        2.1.4 Point 4: Operating Procedures. 23
        2.1.5 Point 5: Employee Training. 23
        2.1.6 Point 6: Contractors 24
        2.1.7 Point 7: Prestart-Up Safety 24
        2.1.8 Point 8: Mechanical Integrity 25
        2.1.9 Point 9: Nonroutine Work (Hot Work) Authorizations. 25
        2.1.10 Point 10: Managing Change 25
        2.1.11 Point 11: Investigating Incidents 26
        2.1.12 Point 12: Emergency Preparedness 26
        2.1.13 Point 13: Compliance Audits 27
        2.1.14 Point 14: Trade Secrets. 27
        2.1.15 Summary: The CST and the 14 Points. 27
    2.2 Common Types of Safety Meetings. 28
    2.3 Common Types of Safety Training 31
        2.3.1 OSHA Regulatory Safety Training 31
        2.3.2 Industry-or Plant-Specific Training. 32
        2.3.3 Safety Programs Evaluation. 33
                Leadership 34
                JSA 34
    2.4 LOTO 34
    2.5 Compliance Documentation. 36
        2.5.1 SDSs. 37
    2.6 Frequently Encountered Safety Equipment 38
    2.7 Start-Up Safety: Practical Examples 39
        2.7.1 PHA 39
        2.7.2 Emergency Drill or Evacuation 40
        2.7.3 Deciding When a Respirator Is Necessary. 40
        2.7.4 Proper LOTO Procedures. 41
        2.7.5 PSSR. 41
    2.8 Safety Instrumented Systems. 41
    Summary. 41
    Review 42
    Recommended Reading. 43

Chapter 3 Documenting the Commissioning and Start-Up Processes. **45**
    3.1 Project Documents. 46
    3.2 Hard Copy and Electronic Documents. 47
        3.2.1 Original (Hard-Copy) Drawings. 47
        3.2.2 Electronic Drawings 48
                Layers. 52
                Lines. 52
                Scale 52

　　　　　　　　File Names, Text, and Notations 52
　　　　　　　　Revisions and Annotations 54
　　　　　　　　Symbols... 54
　　　　　　　　Drawing References.................................. 55
　　　3.3　Document Locations 56
　　　3.4　Commissioning and Start-Up Drawings and Documents........ 57
　　　　　3.4.1　Gantt Charts..................................... 57
　　　　　3.4.2　Functional Specifications 58
　　　　　3.4.3　PFDs.. 59
　　　　　　　　Mass (or Material) and Energy Balance 61
　　　　　　　　Energy Balance...................................... 62
　　　　　3.4.4　P&IDs... 62
　　　　　3.4.5　General Arrangement Drawings and Plot (Location) Plans... 63
　　　　　3.4.6　Instrument Specification Sheets.................... 65
　　　　　3.4.7　Loop Diagrams (Loop Sheets) 67
　　　　　3.4.8　Loop Check Log Sheets............................ 68
　　　　　3.4.9　Calibration Data Sheets........................... 69
　　　　　3.4.10　Installation Details.............................. 69
　　　　　3.4.11　Other Documents and Drawings 70
　　　　　　　　Manufacturers' Information Documents 70
　　　　　　　　Electrical Wiring Diagrams 71
　　　　　　　　Network Topology Diagrams.......................... 73
　　　　　　　　Logic Diagrams and Other Control Logic Documentation.. 73
　　　　　　　　Flowcharts.. 75
　　　　　　　　SOPs... 75
　　　　　　　　Quality/Inspection Checklist........................... 77
　　　　　　　　Start-Up Plan 77
　　　　　　　　Test Documents 77
　　　　　　　　FAT Documents 79
　　　　　　　　SIT Documents..................................... 80
　　　　　　　　SAT Documents 81
　　　　　　　　Process Validation Documents 81
　　　3.5　Documents Used by the CST 83
　　　3.6　Control of Project Documents 87
　　　Summary.. 87
　　　Review... 88
　　　Recommended Reading.. 88

Chapter 4　Working with Others 91
　　　4.1　Start-Up Team Organization 93
　　　　　4.1.1　Operations Department 93
　　　　　4.1.2　Control Systems Engineer.......................... 95
　　　　　4.1.3　Project Engineering Department..................... 96
　　　　　4.1.4　Process Engineering (Technical Representative)........ 96
　　　　　4.1.5　The Project Manager and Other Management Personnel ... 98
　　　　　4.1.6　Maintenance Department........................... 99
　　　　　4.1.7　Environmental Department......................... 99

		Continuous Emission Monitoring Systems 101
		Predictive Emission Monitoring Systems. 102
		Leak Detection and Repair Programs. 102
	4.1.8	Purchasing Department 102
	4.1.9	Human Resources (Personnel). 103
	4.1.10	IT Department 104

4.2 CST Interaction with Other Personnel 106
    4.2.1 Contractors 107
    4.2.2 TPPS Representatives 107
    4.2.3 PAS Vendors and System Integrators 108
    4.2.4 Personnel from Other Plants 109
4.3 Chain of Command 110
Summary 111
Review 111
Recommended Reading. 112

Chapter 5 Verifying and Managing Changes 113
5.1 The MOC Process. 114
5.2 Maintenance, Upkeep, and Control of Project Documents 118
Summary 119
Review 120
Recommended Reading. 120

Chapter 6 Personnel Requirements and Responsibilities 121
6.1 Staffing and Overtime 121
    6.1.1 Division of Responsibility 122
    6.1.2 Task Assignments and Responsibilities 123
    6.1.3 Scheduling 124
    6.1.4 Workload and Priorities 126
    6.1.5 Coverage during Different Start-Up Phases. 127
6.2 Training 127
    6.2.1 Installation Training 128
    6.2.2 Configuration Training 128
    6.2.3 Maintenance Training. 130
    6.2.4 Other Training and Equipment 131
6.3 Emergency Contacts 133
Summary 134
Review 134
Recommended Reading. 135

Chapter 7 Prestart-up Activities 137
7.1 Equipment and Instrumentation Installation 139
7.2 Quality Assurance/Quality Control Inspection 139
7.3 Configuration 141
7.4 Mechanical Completion 142
7.5 Pre-Commissioning 142
7.6 SIMOPS 144

	7.7	CST Tools, Test Equipment, and Technology 145
	7.8	Working with Vendor Representatives and Specialists 146
	7.9	Commissioning 147
	Summary... 152
	Review 152
	Recommended Reading................................... 154

Chapter 8   The Start-Up... 155
    8.1 Start-Up Overview..................................... 155
    8.2 PSSR.. 160
    8.3 Start-Up Plan... 165
        8.3.1 Staffing 166
        8.3.2 System Start-up Order........................... 167
        8.3.3 Introducing Feedstock........................... 168
        8.3.4 Initial Ramp-up 169
        8.3.5 Emergency Shutdown Testing 169
        8.3.6 Loop Tuning 170
        8.3.7 Environmental Testing 172
    8.4 Issues and Problems 173
        8.4.1 Environmental Problems 173
        8.4.2 Processing Problems 173
        8.4.3 Equipment Problems............................ 173
    8.5 Final Acceptance Audit................................. 173
    8.6 Operation and Post-Project Activities 174
        8.6.1 Sustainable Steady-State Operation..................174
        8.6.2 Project Close-Out 175
    Summary..176
    Conclusion..176
    Postscript: ISA CAP 176
    Review 177
    Recommended Reading................................... 177

Appendix A: Instrument Repair Technician Job Description 179

Appendix B: Safety Data Sheet (SDS)................................... 183

Appendix C: Factory Acceptance Testing Checklist 195

Appendix D: Answers to Review Questions............................. 205

Appendix E: ISA Form 20.50 229

Appendix F: Acronyms 231

Index 237

About the Author

Born in New York City, Diane R. Barkin graduated with a BS in chemical engineering from the University of Pittsburgh. After spending a decade in manufacturing process operations, she worked full time in the automation of industrial processes and started her own automation consulting company, Harris Automation Services, Inc., in 1999.

Barkin has worked as a project manager, team lead, and programmer for automation projects from conception, design, and specification through award, programming, training, start-up, and maintenance.

Barkin has extensive experience with automation document and distributed control system (DCS) development, specification, bid and reward activities, as well as with programming and start-up, as-built documentation, and providing complete packages for clients and engineering firms.

Barkin's DCS experience includes Honeywell (Legacy and Experion^{PKS}), Foxboro I/A, and Emerson DeltaV, including programmable logic controller (PLC) interfaces to all three of these products.

1

The Role of the Control Systems Technician

Project Execution and Control (Project Management) Overview
Job Descriptions
CST Roles
CST Tasks

In this book, a *start-up* refers to the project work associated with starting up a newly built (greenfield), substantially changed (brownfield) manufacturing facility, or the start-up of a portion of a manufacturing facility that has had a process change requiring new equipment or controls.

The responsibilities of the control systems technician (CST)[1] before and during start-up include installation, troubleshooting, repair, and routine maintenance of instrumentation and controls. A large portion of this work includes troubleshooting and coordinating repairs to control loops throughout the plant. This may also include working on the control systems that connect the instrumentation and process equipment via hardware and software to enable operators to run the plant from industrial computers.

The CST follows company and craft standards and specifications, and coordinates his or her activities with Operations and with other crafts and departments. A CST may be a direct or contract employee.

---

1 A CST may also be called an *instrument and electrical* (I&E) *technician*.

As a CST, this may be the first time you have been asked to participate in start-up activities; or perhaps you are very experienced and have participated in many start-ups during your career. In any event, as part of the commissioning and start-up team, you should know what your roles and responsibilities are during this critical period. The chapters of this book cover the major aspects of a CST's involvement in a plant's start-up activities:

- Chapter 1, "The Role of the Control Systems Technician"
- Chapter 2, "Applicable Safety Practices and Standards"
- Chapter 3, "Documenting the Commissioning and Start-Up Processes"
- Chapter 4, "Working with Others"
- Chapter 5, "Verifying and Managing Changes"
- Chapter 6, "Personnel Requirements and Responsibilities"
- Chapter 7, "Prestart-up Activities"
- Chapter 8, "The Start-Up"

Sometimes start-ups must be accomplished with limited personnel resources and under extreme time pressure. An effective start-up plan, which is discussed in Chapter 8, depends on knowing what CSTs should do and how many CSTs will be required—and available—to do it. Management will take personnel limitations into account, but CSTs also must understand the roles of the personnel involved in the start-up. Sometimes this information is communicated verbally, and sometimes in writing in formal job descriptions. This chapter will discuss the importance of job descriptions, and the different roles and tasks a CST may be involved with throughout the project leading up to the start-up.

The role of the CST varies from plant to plant. For example, plant size may determine the roles a CST will play. If the CST is part of a small group, or possibly working alone, the job will be different than that of a CST working with a larger department. In a larger organization, the CST may have less independence and specific jobs he or she has to perform during start-up; the CST will not necessarily be involved in every function described in this book. Conversely, a CST working in a small facility may be responsible for many, if not all, tasks described in this book.

The information in this book is intended to apply to many different types of industries that might employ a CST. Obviously, the type of industry, the size of the plant and/or project, and management will determine how the CST will function within the organization and during start-up.

Project Execution and Control (Project Management) Overview

The start-up process cannot be addressed without discussing the steps and deliverables necessary to complete the work to successfully start up the new and/or modified facility. Some of the work the CST is responsible for occurs during preparation for the start-up, such as during construction (instrument installation), pre-commissioning, and commissioning (loop and function checks).

Project execution and control (project management) are at the heart of a project. Planning, organizing, securing, and managing resources is necessary for the safe and successful completion of specific project goals and objectives. Resources include people, materials, and money. Project goals may include a set of agreed-on project deliverables. Examples of project deliverables are:

- **Plans** – Construction, procedures, and objectives
- **Drawings** – As-built drawings, loop sheets, documents, and loop folders
- **Calculations** – Studies, material, and energy balances
- **Purchased equipment** – Tanks/vessels, instrumentation, control systems, and copper and fiber-optic cabling
- **Installation services** – Installation, inspection, initialization (power-up), loop and function checking, programming, training, factory acceptance testing (FAT), and site acceptance testing (SAT)

A project has several steps that are common throughout industries. These steps progress as money is approved by the "owner" company. Sometimes a project is cancelled during or after a phase if the project is not deemed feasible or other factors come into play (environmental permits are not gained, cost cannot be justified, etc.). Common phases of a project from inception to start-up include the following:

- **Front end engineering and design (FEED)** – A project is proposed, and budgetary design is performed to determine whether the project is feasible (i.e., if there is a good return on investment for the company).
- **Detailed design** – The project has been approved for funding. Personnel are engaged to perform engineering design for all systems, and documentation packages are prepared for the client's review and subsequent equipment purchase and construction.
- **FAT** – Control systems have been designed and purchased. Software has been completed during detailed design and installed on the purchased system

(normally in a staging area of the firm that did the programming). A team of representatives for the plant—for example, operators, CSTs, and engineering personnel—check the system hardware and software, and sign off their acceptance of the system before it is shipped to the plant.

- **Construction** – Instrumentation, equipment (e.g., vessels and pumps) and supporting structures (e.g., cable trays, wiring, and grounding) are installed. The project is deemed "mechanically complete."

- **Pre-commissioning** – This includes activities that involve the introduction of fluids—but not hydrocarbons—into systems for cleaning (e.g., nitrogen blows), instrument loop and panel function checks, the energizing of electrical equipment, and the running of motors without loads.

- **Commissioning** – Logic function checks are performed—including safety, logic, and interlock testing—then equipment and systems are brought into operation safely and in the correct sequence.

- **Start-up** – After a pre-startup safety review (PSSR) is performed, the plant is ready for the initial introduction of hydrocarbons (where applicable) and the process is ramped up to full production.

If these phases overlap, conflicts occur, most commonly between construction and pre-commissioning activities. This is referred to as simultaneous operations (SIMOPS) and is discussed in Chapter 7.

1.1 Job Descriptions

The job description not only defines work responsibilities for the person doing the job, but it also helps the employer explain what the position requires when hiring and during performance evaluations. As a CST, your job description will cover your essential job functions, including those required during a plant start-up. Management can use the job description when hiring to review the position with a potential employee, orient new employees to their positions, and evaluate current employees' job performance. Some job descriptions are very specific, whereas others are more general to allow for flexibility in work assignment. Job descriptions include the following elements:

- Job summary statement
- Basic functions of the position
- Responsibilities of the position (identifying essential tasks)

- Supervisory responsibilities (if applicable)
- Skills necessary for the position (e.g., an understanding of math as it pertains to electrical and instrument equipment, computer use and applications, electrical theory, and instrument and associated equipment installation)
- Experience necessary for the position
- Work environment (including physical surroundings)
- Education requirements (e.g., trade school, formal training, or other certifications, such as an International Society of Automation—ISA Certified Control Systems Technician—CCST)
- To whom the position reports (organizational relationships)
- Date of the most recent update to the job description
- Who prepared the job description
- Who approved the job description

Relevant Standards

American National Standards Institute/International Society of Automation

- ANSI/ISA-98.00.01-2002, *Qualifications and Certification of Control System Technicians* (withdrawn in 2014, listed for historical purposes)

International Society of Automation

- ISA-67.14.01-2000, *Qualifications and Certification of Instrumentation and Control Technicians in Nuclear Facilities*

As shown in Appendix A, typical job elements that might pertain to project and start- up roles for a CST include the following:

- **Essential functions** – Calibration, troubleshooting, and inspection of newly installed equipment including process automation systems (PASs), which include distributed control systems (DCSs), programmable logic controllers (PLCs), safety instrumented systems (SISs), other systems that interface to the PAS, and computers and software applications such as Microsoft Office. Optional essential functions, depending on the CST's roles, may be knowledge of project management software (e.g., Microsoft Project) and enterprise software (e.g., Systems, Applications, and Products in Data Processing—SAP). The relationship between the PAS and higher level computer systems and their software is discussed in Chapter 4 (ANSI/ISA 95 Model). Additional essential functions might require the CST to periodically work with a start-up team and to work shifts.

- **Behavioral capacities required to perform essential functions** – Safety awareness; technical judgment; conscientiousness; initiative; adherence to policies and procedures; maturity; teamwork, interpersonal, and organizational skills; and the ability to work odd, and sometimes long, hours and to work with different disciplines and crafts (e.g., electricians, pipe fitters, engineers, contractors, and operators).

- **Physical requirements** – The ability to work in environments with loud noises, extreme temperatures, and extreme ambient conditions (e.g., dust); the ability to work while wearing personal protective equipment (PPE), such as gloves and respirators; and the ability to climb to heights using ladders and manlifts, where necessary.

In addition to these points, the following may be described in company job descriptions:

- **Supervisory and nonsupervisory responsibilities** – These include the tasks associated with a start-up, such as calibration and loop checking, and the need to use special tools. The job description might also state that the CST could have direct reports.

- **Experience required** – This element might indicate that the CST should have a certain number of years on the job to qualify for start-up involvement, specifically including start-up experience. (This is not necessarily a catch–22, but it may be how managers differentiate between multiple candidates for the same job.) CSTs often gain start-up experience by following an experienced "hand" before they are allowed to be a "lead tech."

1.2 CST Roles

The CST's roles before and during start-up include but are not limited to the following:

- Trainer or instructor
- Safety inspector
- Vendor and service representative or specialist liaison

- Technician performing installation, calibration, loop checking, programming, or troubleshooting
- Quality inspector (QA/QC)
- Assistant to the Design and Engineering departments
- Leader

1.2.1 Training

As an experienced CST, you may be asked to train and instruct others in activities associated with the project and subsequent start-up, as well as in ways to maintain the new plant once it is online. You might be asked to instruct people in programming the PAS, using handheld communicators and process calibrators, or performing loop checks according to the plant standard. You could be considered for this role based on your experience, your communication skills, and other important attributes such as patience and the ability to teach.

A company benefits from using in-house staff to conduct this type of training because when people teach, they also learn. When you instruct someone else, you not only help that person understand but you also reinforce your own knowledge. Additionally, a teacher who is weak in one aspect of the subject will most likely study the material before attempting to teach it, thus learning more in the process. As a CST, you should undertake the challenging role of trainer if your company offers it.

1.2.2 Safety Inspector

Safety is discussed at length in Chapter 2 because a start-up tends to be more hazardous than normal operation. This is due to time constraints; construction activities, such as temporary power, scaffolding, trenches, and systems that are not in operation yet; pressure; and inexperience—especially in a new process. Everyone at a plant is responsible for safety, but at times you may be asked to serve as an inspector or a safety team leader during and after the start-up. Because CSTs are knowledgeable about instrumentation, they make excellent safety inspectors and safety team leaders. If asked to participate in a plant safety team, the CST provides a different perspective from other members who may be from other disciplines. The CST may also be asked to participate in a process hazard analysis (PHA), an activity described in Chapter 2.

If an unsafe act or accident occurs, as a CST you may be interviewed and asked about the incident or be asked to interview others during a safety audit or investigation. Even if you are not assigned a specific safety role, like everyone else, you should practice and help enforce company safety rules for every task you are assigned.

Safety training is a requirement, and the CST must participate in (refresher) safety training on a regular basis. A training curriculum is detailed in Chapter 2 and includes the following:

- The Occupational Safety and Health Administration (OSHA) process safety management (PSM) standard
- The OSHA hazard communication standard
- Lockout/tagout
- Nonroutine maintenance
- PPE

It is the CST's responsibility to receive training and to participate in, and sometimes lead, training efforts to maintain high safety standards during and after plant start-up. Any activities you are involved with must adhere to all safety rules and procedures required by the company and by the state and federal government.

Relevant Standards

US Occupational Safety and Health Administration (OSHA)
- 29 CFR 1910.119, *Process Safety Management of Highly Hazardous Chemicals*, July 1, 2010
- 29 CFR 1910.1200, *Hazard Communication*, May 2012

1.2.3 Liaison

Just before start-up, a facility's management may summon expert help from equipment vendors, PAS suppliers, and other specialists. Normally, these are vendor representatives (referred to in this section as "specialists") who are paid a high hourly rate to come to the plant site to work with plant personnel. When vendors or specialists enter a plant, as a CST you may be directed to help them complete their work, as well as serve as a liaison between them and installation contractors—especially when their work involves instrumentation and controls. Often vendor representatives know only their own small part of a plant or the piece of equipment they make, so you must ensure that their part works properly with the rest of the plant's equipment, including instrumentation.

You may also be asked to learn from one of these specialists during the start-up period so you can later instruct plant employees. For example, a specialist who is helping to start up a weighing system may be very familiar with the software that controls it. This is your opportunity to learn both the system and the software, as well as how to troubleshoot it, from the most knowledgeable person. A CST should take advantage of such opportunities by asking questions, taking notes, and doing hands-on work alongside an expert.

While working with vendors and specialists, the CST normally must act as an escort, staying with them the entire time they are in the plant. These specialists might need an escort because they are restricted from hazardous areas and are not required to wear the same safety equipment that plant workers wear. An escort might also be necessary because the specialists are less familiar with the plant or are not allowed to see proprietary areas or equipment. It will be up to the escort to ensure that the specialists follow plant rules and that they are aided during their visit so they may still perform their jobs effectively.

The escort must ensure that the specialists follow the company's safety rules and good maintenance practices, and that what they do is documented (e.g., sign work permits, where applicable, and redline documents to reflect changes in their systems, such as wiring diagrams). As these specialists are often paid a premium, their time must be used efficiently. The company must be spared additional expenses, and the specialists must complete their jobs effectively so nonperforming equipment does not delay start-up.

Contractors are a different story. Though you may also be assigned to work with them when they install equipment, normally you do not have to serve as an escort. Contractors remain onsite for a long time and therefore must go through the same safety training and wear the same safety equipment as plant operations personnel. They are responsible for receiving training through their contract company. They must follow the same rules as all plant personnel. Much of their work, such as installing large pieces of equipment like tanks, does not require a CST's attention until an interface to instrumentation and controls is required. Then, when the vessels are completed, you as the CST will communicate with the contractors and plant management to coordinate and begin installation and check-out of associated instrumentation.

If you observe failures of contractor personnel or others to follow safety or maintenance procedures, work must cease until the incidents have been reported and corrective actions have been taken.

1.2.4 Technician Duties

As a CST, your role during the project will also include most of the traditional duties of a technician: installation, calibration, loop and function checking, programming, and troubleshooting, which are discussed elsewhere in this book and in the other ISA Technician Series books listed in the "Recommended Reading" section in Chapter 7.

Additional duties may include PAS configuration and programming (e.g., input/output—I/O, graphics, and trends) or modification of control schemes (e.g., range change in field and on PAS and I/O assignment modification), if you have these skills.

1.2.5 Inspection (QA/QC)

Work quality is always important, and any work that a CST performs should be of a high caliber. Prior to a start-up, CSTs and other technicians may be assigned quality/inspection jobs as well as other plant activities. This is often referred to as quality assurance/quality control (QA/QC). For a plant start-up to be successful, good organization is as important as workmanship. Though people should be self- governing and check the quality of their own work, sometimes a second set of eyes is needed. You may or may not be given the title "quality inspector," but you will nevertheless be expected to function as an inspector. Remember that the workmanship of contract or vendor personnel can directly affect your downstream work if you are the CST assigned to the facility or the lead technician for other CSTs. For example, loose wiring or unlabeled wiring can make troubleshooting or locating loops difficult. Typically, for the CST, quality inspections focus on housekeeping, wiring runs and terminations, wiring tags, instrumentation, PAS, and documentation.

Sometimes the project is too big for a CST to function as a quality inspector and perform the normal CST tasks (installation, calibration, loop check, etc.). In this case, the CST may work as one or the other or a separate quality/inspection team will be set up (by craft) using either plant personnel (if there are enough to support plant construction, inspection, and start-up activities) or a quality/inspection contracting company dedicated entirely to inspections. A third-party quality/inspection contracting company that was not involved with the construction and is not the end user (plant owner) may be less biased and more critical, therefore they may perform very strict inspections that result in a well-executed project and successful start-up.

Housekeeping

Housekeeping means what it sounds like: ensuring that CSTs, subordinates, and any vendor specialists or contractors involved in the start-up keep their areas clean and discard waste materials daily. Proper disposal of packaging materials from wiring and instrumentation, pieces of wiring and tubing, tie wraps (used to secure wiring in bundles), and out-of-date paperwork and documents must occur. Doing this is particularly important for trash on the floor as it can create a fall or fire hazard. Out-of-date paperwork should be disposed of to avoid confusion and enhance organization.

Wiring Runs and Terminations

Wiring runs and terminations run between field instrumentation (i.e., located in the process areas), junction boxes, marshalling panels, the PAS, and the control room. These cables should be run properly and neatly in cable trays and conduits. Tray covers and all equipment should be closed ("buttoned-up") before start-up begins; if wiring runs travel through areas designated with "hazardous" classifications (e.g., a Class I, Division 1 or Class I, Division 2 area), appropriate seals must be poured after the connections are complete and loop checked. Cables carrying different voltage levels should be run in separate trays, with each cable lying neatly in parallel.

Relevant Recommended Practice

American National Standards Institute/International Society of Automation

- ISA-RP12.06.01-2003, *Recommended Practice for Wiring Methods for Hazardous (Classified) Locations Instrumentation – Part 1: Intrinsic Safety*

Wiring Tags

Because there are so many connections and wires, several people will be needed to ensure that the work is complete and performed accurately. Junction boxes, which connect "home run" cables from equipment in the field to the PAS, contain a lot of wiring that should be terminated and labeled at both ends. Unused wire pairs should be terminated in terminal strips. "Taping off" is bad practice. Everything in these cabinets should be documented on electrical and other drawings, as well as labeled with wiring tags. Discrepancies should be noted and redlined in the engineering documentation by the installer and/or inspector and the marked-up copy given to a draftsperson or electrical designer to be updated. These Management of Change (MOC) procedures will be discussed in Chapters 3 and 5.

As an inspector, a CST must examine wire tags and labeling to ensure consistency and accuracy. The labels should be printed legibly and secured to the wire neatly so that wire designations can be easily read. Quality work in this area will pay off in the future because well-labeled wiring runs make it much easier to troubleshoot and maintain loops.

Instrumentation

There are many ways a CST may be involved with the instrumentation required for the project: replacement procurement, if needed; installation; calibration; and loop checking. High-quality work is required at all times. If the CST is involved with procuring instruments and associated equipment, it is important that the CST understands the process requirements and ensures that the correct equipment is ordered for the installation. Important factors to consider are process characteristics such as ambient and operating pressures and temperatures, pH, corrosive or abrasive environments, required material of construction, and the type of application for which the instrumentation is to be used. If the CST has this information, then the documentation used for instrument specification (often called a *spec* or *data sheet*) and the purchase will be accurate, and the proper equipment will be installed.

It is also important that the CST fills out data sheets correctly. If the CST is involved with ordering or reordering this type of equipment (due to failure or changes), he or she should be clear about specifying the correct equipment and should ensure that all information (e.g., model number) matches the requirement and the spec sheet. As a CST involved with instrumentation, it is important to work safely and efficiently and to use your expertise and experience to perform a quality job.

See Chapter 3, which discusses instrument spec sheets and provides the relevant ISA standards associated with instrument specification forms.

PASs

If you are involved with making configuration or program changes to a PAS, including the DCS/basic process control system (BPCS), PLC, or any other programmable device, then a quality job is as essential as a safe one. Once you are assigned the job of making a change and determining the optimum time to make this change (normally controlled by the plant MOC procedure), it is important to communicate with those who will be affected by the change. Considering whether the change can be done online or offline is critical. While making the change, check your work, document what you have done by embedding this information in the program or in a software "field," and test the change before implementation. When you are ready to implement the change (e.g., a download to a process controller), alerting the proper people and ensuring safety are vital. You should be fully qualified to work on the equipment. In addition to MOC requirements, you should also do the following when making any PAS changes: include revision information with software and document changes, provide your initials or signature as the one making the change, update any related documentation or procedures, return the equipment to service, and verify that the process is reacting or performing properly after the modification.

Documentation

Documentation is a critical part of the design, start-up, and future maintenance of a plant. Therefore, the documentation must be examined for accuracy and revised as necessary to maintain an accurate representation of the plant. It must reflect the plant, even after changes have been made to the process, equipment, and instrumentation. To document this as built status, as discussed in Chapter 3, the CST must depend on his or her knowledge of process control terminology and experience. Being involved with the start-up will make the CST a valuable reviewer of such materials.

1.2.6 Design and Engineering Assistance

As a CST, you might become involved with the project during the plant design phase. This involvement may include contributing to design changes during PHAs, as discussed in Chapter 2. You can leverage your experience and expertise to generate or answer questions during PHA what-if discussions. For example, you may notice the potential for a hazardous release if a certain type of control loop or interlock strategy is not employed. Your explanations of the problem will help convince people of the danger or help them understand why additional instrumentation or control is necessary. In this way, you can be part of the process, helping to design control schemes to minimize or prevent accidents. In addition, a CST can offer design and engineering expertise in the areas of maintainability and constructability. Proper mechanical design of instrument loops helps to ensure an efficient and safe start-up, ease in (future) maintenance, and reliability. For example, it is often the CST who needs access to transmitters that may initially be placed in an inaccessible location. The CST can offer suggestions to avoid these types of mistakes.

Engineering firms need contact with plant personnel, including CSTs, to ensure that their design and ideas will work. You must be open and honest when solicited for your opinion. Your knowledge will be respected, and you may learn something too.

1.2.7 Leader

The CST may have other technicians working with and for him or her. As a leader, the CST is expected to coordinate these people's activities, handle associated paperwork, and attend project review and production status meetings to discuss all work that is in progress and completed by his or her people.

When the CST is a leader, he or she functions as a teacher and should therefore communicate well and know how to handle personnel. In this role, the CST must be able to lead and instruct people as necessary, as well as perform the other technical tasks in the CST's job description. In addition, CSTs who are leaders are responsible for their team's safety, and possibly their work schedule and any tools or materials they might require.

1.3 CST Tasks

Baseline tasks are the basic duties and responsibilities a CST performs each day of the project and in preparation for the start-up. Examples of these are installation, calibration, and loop and function checks. Built on this base may be other duties, as described in the previous section, that have shorter time spans and occur at random intervals. Examples of these include participating in PHA and escorting vendor specialists. Additional duties may occur later during the start-up process, such as compiling loop folders and providing as-built drawings and document mark-ups.

1.3.1 Baseline Duties

Examples of baseline duties are as follows:

- Checking the work order system and plant logbooks
- Talking with the operators about problems or instrumentation issues that must be dealt with immediately
- Monitoring the progress of calibration and loop checking (according to schedule)
- Performing daily record keeping
- Cleaning up the work area after the workday is complete

High-priority baseline activities should be dealt with first. Assuming that people on the night shift were performing calibrations and checking loops, a supervisory CST's first baseline activity in the morning normally would be to check the progress of the night-shift employees' work, and then to check throughout the day until loop checking is complete.

Record keeping and housekeeping should be done every day and be completed by the end of each shift. Some of the daily tasks a CST will be involved with are calibration, verification of calibration, and loop checking. One plant surveyed for this book verified calibration and performed loop checking at the same time.

Given the complexity of the systems, the CST working a start-up is typically scheduled to work 6 or 7 days a week until the start-up is complete. Additionally, personnel are expected to be flexible because start-ups can be somewhat unpredictable, especially if there are difficulties with getting something to work. Technicians are often scheduled to work for 10 to 12 hours a day. If necessary, they work longer to complete a task, although companies rarely allow technicians to work longer than 18 hours a day to ensure that employees' (and others') safety is not compromised. Depending on many factors, including the size of the plant, these baseline start-up tasks can go on for many weeks—or even months. Along with serving as a liaison, expert, and teacher or instructor, these baseline activities are a major part of the CST's daily routine.

Summary

A CST plays an important part in the start-up of a new facility. You should know what your roles and responsibilities are so you can perform your job effectively and safely. Depending on the plant and your experience, you may perform baseline duties as well as serve as a leader and instructor. You may also assist in design and engineering. The duration of baseline duties will depend on the size of the plant and how quickly required start-up tasks are successfully completed. As Chapter 4 will explain further, all personnel working in the start-up should understand the extent of their involvement and contribute in their areas of expertise, working with several disciplines. In the long run, informed, skilled people and effective teamwork will help the new plant to start up safely and successfully.

Review

1.1 With regard to wiring, what types of high-quality work must a CST perform?

1.2 Give examples of situations in which a CST may serve as an assistant during plant design.

1.3 What would you consider to be essential tasks for a CST during start-up?

1.4 Why can a CST function as an effective safety and quality inspector?

1.5 How do vendors get involved during the start-up?

1.6 What are some purposes of a job description?

1.7 Who might report to a lead CST, and what are a lead CST's responsibilities?

1.8 What does a CST do when serving as a liaison between vendors and/or contractors and the other members of the start-up team or plant organization?

1.9 What level of education and training do you think a CST should have in order to be involved in a start-up?

1.10 What is meant by *baseline work*?

1.11 In which subjects do you think a CST could teach, train, or instruct people?

1.12 Why is it important for a CST to communicate frequently with vendor representatives and specialists?

Recommended Reading

Berge, Jonas. *Software for Automation: Architecture, Integration, and Security*. Research Triangle Park, NC: ISA (International Society of Automation), 2005.

Lipták, Béla G., ed. *Instrument Engineers' Handbook. Vol. 1, Process Measurement and Analysis*. 4th ed. Boca Raton, FL: CRC Press/ISA (International Society of Automation), 2003.

Murrill, P. W. *Fundamentals of Process Control Theory*. 3rd ed. Research Triangle Park, NC: ISA (International Society of Automation), 2000.

Sherman, R. E., and L. Rhodes, eds. *Analytical Instrumentation*. Research Triangle Park, NC: ISA (International Society of Automation), 1996.

Staples, Leo. *Project Management: A Technician's Guide*. Research Triangle Park, NC: ISA (International Society of Automation), 2010

2
Applicable Safety Practices and Standards

The 14 Points of OSHA PSM
Common Types of Safety Meetings
Common Types of Safety Training
LOTO
Compliance Documentation
Frequently Encountered Safety Equipment
Start-Up Safety: Practical Examples
Safety Instrumented Systems

The project we are referring to in this book involves many factors that contribute to the need for increased safety awareness. Typical hazards include the following:

- An increased number of people in the plant area
- Production deadlines contributing to emotional stress for involved personnel
- New people or new processes introduced into the plant area
- Difficulty in getting new equipment running and checked out
- Human error, including errors in installation, design, configuration, and judgment

Certain sites may face additional hazards, including:

- The introduction of hazardous materials to the facility for the first time
- The possibility of pressure and/or temperature extremes
- Machinery that can harm plant personnel if not used carefully

It is important to train personnel in relevant emergency preparedness and to ensure that their training meets the safety needs for processes, such as process shutdown, emergencies, conservatism during the first operating stages, troubleshooting, identification of small problems, communication of needs, and management of "hot work" (nonroutine or unplanned work) and emergency repairs.

Before start-up, the construction manager (CM) takes the lead on safety, typically working with an individual or a team assigned to safety. Often, a Certified Safety Professional (CSP) is either the leader of or a member of the team. The CM is responsible for working with Operations personnel while maintaining a safe work environment and providing safety training prior to allowing anyone to work on the project site.

Before start-up, management should provide training and have a plan for the construction, commissioning, and starting up phases of the project. Particular scrutiny should be given to lockout/tagout (LOTO) procedures, behaviors during the project, potential equipment malfunctions, and potential design errors. Specifics pertaining to the plant, such as pressure relief valves, high- and low-temperature units, pressure or level cutouts, hardwired and software interlocks, burner management systems (BMSs), and other safety systems must be documented and understood. Management should require many meetings before a plant is started up, beginning months, even years, in advance. As start-up gets closer, these meetings will become more frequent, for example, weekly or even daily.

This chapter will describe and give examples of safety practices and standards used in manufacturing. All manufacturing facilities have both *greenfield* (completely new) start-ups and *partial* start-ups (start-ups resulting from plant or process automation system—PAS modifications, replacements, and upgrades, as well as additions to existing processes). The practices and standards discussed in this chapter apply to most employees at manufacturing facilities and span many sectors, including chemical, oil and gas, public and private utility, pulp and paper, food, and pharmaceutical industries. Although much of this chapter applies to all manufacturing and support personnel, the discussion on these topics will be expanded to show how they apply to the work of a CST. The information presented affects a CST's job, and it is important to understand how it does so.

As mentioned in Chapter 1, the process of preparing for a start-up is a busy time when a great deal of work must be accomplished and the number of people available to help is limited. The concepts discussed in this chapter may therefore be considered ideals that may not always be practiced and adhered to in the "real world" (though to a large extent they are), especially during commissioning and start-up. The realities of the project may make it necessary to deviate from these practices to get parts of a plant running and equipment working; however, US Occupational Safety and Health Administration (OSHA) standards and potential penalties still apply during this time. It is important that you recognize these deviations, understand why they are necessary, and help rectify the situation quickly, effectively, and safely.

Applicable safety practices, standards, and equipment addressed in this chapter are as follows:

- OSHA process safety management (PSM) standard
- Safety training
- LOTO
- Compliance documentation
- Safety equipment
- Start-up safety: practical examples
- Safety instrumented systems (SISs)

The OSHA PSM standard is included here because PSM procedures affect most plant employees and impact plant safety at all times, including during start-up. Plants with highly hazardous materials (and even many of those without them) have standardized their procedures and operations based on OSHA PSM. The following 14 points of PSM are described in the next section:

1. Employee involvement
2. Process safety information
3. Process hazard analysis (PHA)

4. Operating procedures
5. Employee training
6. Contractors
7. Prestart-up safety
8. Mechanical integrity
9. Nonroutine (hot) work authorizations
10. Managing change
11. Investigating incidents
12. Emergency preparedness
13. Compliance audits
14. Trade secrets

Most facilities are regulated by several entities or organizations. Depending on the country and industry, these include the Food and Drug Administration (FDA), which uses quality systems regulations (QSRs), formerly known as good manufacturing practices (GMPs); the International Standards Organization (ISO); the International Electrotechnical Commission (IEC); and OSHA. In addition to regulations from these organizations, individual states and countries have regulations that industries must meet.

Sites seeking OSHA Star certification must follow a set of rules. Sites awarded OSHA Star certifications are recognized as complying with OSHA regulations and having an ongoing compliance program. They may be self-regulating and maintain records for their customers, who audit the facilities on a regular basis. OSHA PSM helps facilities assemble the paperwork to pass such audits.

2.1 The 14 Points of OSHA PSM

In 1992, OSHA published *Process Safety Management of Highly Hazardous Chemicals*. This standard is one of many general and permanent rules published in the US *Federal Register* as part of the Code of Federal Regulations (CFR). This section concentrates primarily on Title 29, Labor, where the OSHA PSM standard is located.

In response to OSHA PSM, many US industries have modified and standardized their practices and procedures. These practices and procedures are used in the everyday operation of facilities but are also relevant to start-up because they must be adhered to during all aspects of plant operation. They are also relevant because work associated with start-up can be particularly hazardous.

An effective PSM program requires a systematic approach to evaluating the whole process. The following 14 points of OSHA Standard 1910.119 address many of the most important elements of this evaluation.

2.1.1 Point 1: Employee Involvement

According to the OSHA standard[1], employees and their representatives must be provided with access to PHA information and other information. One way to accomplish this is to make the information available in control rooms and through safety meetings and safety training. As the CST involved in a new plant start-up, you will receive training on the new process and potential hazards. Although much safety information and training are available, it is always the employees' responsibility to absorb the training, review the safety material, and apply safe practices as a continual part of their jobs.[2]

2.1.2 Point 2: Process Safety Information

Complete written information concerning process chemicals, technology, and equipment is essential for an effective PSM program as well as for performing a PHA.

Process safety information can come from many sources:

- **Safety Data Sheets (SDSs)**[3] – These are discussed in Chapter 3.

- **Process technology information** – For example, block flow and process flow diagrams help employees understand the process. These drawings will be discussed and examples given in Chapter 3.

- **Good engineering practice** – For example, information pertaining to process equipment design is documented and references the required codes and standards often provided by the organizations cited below.

---

1 "Occupational Safety and Health Administration," OSHA, accessed February 25, 2019, www.osha.gov.
2 "Employer News Release," US Department of Labor, last modified November 8, 2018, http:// www.bls.gov/news.release/osh.toc.htm.
3 Other entities may refer to this type of document as a *Material Safety Data Sheet* (MSDS) or a *Product Safety Data Sheet* (PSDS).

Many organizations issue standards, specifications, recommended practices, and technical reports that impact process design. These are often referred to as *industry practices*. Issuing organizations include the following:

- American Petroleum Institute (API)
- American National Standards Institute (ANSI)
- Institute of Electrical and Electronics Engineers (IEEE)
- International Society of Automation (ISA)
- National Fire Protection Association (NFPA)

These organizations' documents are cited in this book where applicable. A CST training to become a Certified Control Systems Technician (CCST) should review industry practices. If your company has access to the information handling services (IHS) website, which allows access to and purchase of many industry standards, then you can review many standards online.

Relevant Internet Reference

IHS Markit
- IHS Markit – https://ihsmarkit.com/index.html

2.1.3 Point 3: PHA

A PHA (sometimes called a *process hazard evaluation*) is a systematic effort to identify and analyze the significance of potential hazards. It focuses on equipment, instrumentation, utilities, human actions (routine and nonroutine), and external factors that might impact a given process. These considerations help determine the process's hazards and potential failure points or failure modes.

When a formal PHA is not done, plant personnel from various disciplines typically establish a series of what-if scenarios involving potential problems, develop solutions that will alleviate them, and set implementation priorities. What-ifs are typically less structured than a PHA and are commonly used for small projects, project changes, and when a Management of Change (MOC) process is required. When more structured methods are necessary, PHA or Layer of Protection Analysis (LOPA) may be applied. An experienced CST with knowledge of PASs, electrical and field (plant) instrumentation, and wiring may be called on to be an expert member of a team performing such analyses. Typical drawings used during a formal or informal PHA are piping and instrumentation drawings (P&IDs) and job safety analyses (JSAs).

Detailed guidance on the content and application of PHA methodologies is available from the American Institute of Chemical Engineers (AIChE) Center for Chemical Process Safety (CCPS).

Relevant Internet Reference

American Institute of Chemical Engineers
- AIChE – http://www.aiche.org/ccps

2.1.4 Point 4: Operating Procedures

Under OSHA PSM, the employer must develop and implement written operating procedures, consistent with the process safety information, that provide clear instructions for safely conducting the activities involved in each process. These procedures should be reviewed to ensure they are accurate and provide employees with practical instructions on how to carry out job duties safely. Because of the CST's knowledge and frequent use of drawings such as P&IDs (see Chapter 3 for an example), the CST may be a reviewer of such procedures. In addition, the CST and others who work in the Instrument and Control department, an instrument and electrical (I&E) shop, or a similarly named department must have procedures for handling the routine and non-routine tasks they perform or are involved in.

These procedures are also important when training personnel, and they must be updated when there is a change in the process, such as an equipment upgrade. For example, Operations personnel, such as the "board" operator, must maintain communication with technicians or contractors who are performing maintenance or new installation work in the process area. Procedures must be written to convey the hazards of the tasks to both Operations personnel and to the individuals actually performing the tasks. Operations personnel should be informed when the work is completed. As a CST who might be working in the field (i.e., on the plant floor) in conjunction with an operator in the control room, you often will be in a good position to make sure the work has been completed and the plant can be operated safely.

2.1.5 Point 5: Employee Training

All employees must protect themselves, their fellow employees, and the citizens of nearby communities by fully understanding the safety and health hazards of the chemicals they work with. Training conducted in compliance with the OSHA Hazard Communication (also known as *Right to Know*) standard helps to ensure this. As part of this training, employees should familiarize themselves with SDSs. Whether you are a plant employee or a contractor, you should pursue applicable training and you must follow all emergency and safety procedures for routine and nonroutine tasks.

Relevant Standard

US Occupational Safety and Health Administration

- 29 CFR 1926.21, *Safety Training and Education*, July 1, 2017

2.1.6 Point 6: Contractors

When selecting a contractor or contract company, the employer must obtain information about the contractor to evaluate its safety performance, safety programs, job skills, knowledge, and certifications.

A company's CST will often interact with contractors; at minimum, the CST's work has the potential to affect the contractor's and employees' safety. Contract employees often remain in a plant after the plant has commenced start-up. Sometimes this is because of changes in the project scope or because unexpected problems or scheduling difficulties arise. The contractor might be asked to help with these situations, and contract employees might work directly with the CST; it depends on how the job responsibilities are delineated.

2.1.7 Point 7: Prestart-Up Safety

The following tasks should be done before the start-up of a new process:

- P&IDs should be complete and up to date. Any redlines should be picked up. See Chapters 3 and 5, which pertain to redlining documents.
- Operating procedures should be in place.
- Operating staff should be trained to run the process.

Companies should fully evaluate their initial start-up procedures and normal operating procedures as part of the pre-startup safety review (PSSR), discussed in Chapter 8; this will help assure a safe initial start-up. You or some member of your department will likely be involved with these reviews. As a CST, your knowledge of P&IDs, loop drawings, and calibration sheets (and consequently of the new process) allows you to participate in an informed way in any discussions relating to safety, start-up, and normal operation.

2.1.8 Point 8: Mechanical Integrity

Equipment must be designed, constructed, installed, and maintained to minimize the risk of release of hazardous chemicals or energy.

As part of the mechanical integrity program, you may be given a work order or assignment that requires you to inspect and repair instrumentation. Doing this is often part of a preventive maintenance program.

A mechanical integrity program demands that equipment installation jobs be properly inspected in the field to ensure that proper materials are used, procedures are followed, equipment is installed per the engineering drawings, codes and/or company engineering and installation practices are followed, and qualified craftsmen are doing the work. The CST will be responsible for ensuring that equipment components (e.g., a valve seat) are replaced with components that are properly constructed and composed of materials that are compatible with the chemical service (materials of construction); an inferior part may not work and may compromise safety. This type of inspection applies to new equipment during start-up and replacement parts after start-up and during turnarounds as equipment fails and must be replaced.

2.1.9 Point 9: Nonroutine Work (Hot Work) Authorizations

Nonroutine work authorization notices or permit procedures describe the steps that maintenance supervisors, contractor representatives, or other authorized personnel should follow to obtain the necessary clearance (permit) to get a job started.

As a CST, you will work on jobs that require work permits and may have to sign the permits and/or LOTO forms. For the safety of everyone in the plant, it is of the utmost importance that you, and others, follow these procedures and communicate with each other throughout the job.

2.1.10 Point 10: Managing Change

- **OSHA 29 CFR 1910.119 definition** – Change includes all modifications to a process and excludes "replacement in kind."

A CST who is involved with a change should make certain that an MOC form has been initiated if anything other than the original part is being replaced or installed. By having intimate knowledge of the MOC process you will ensure that your involvement in modifications will maintain plant safety.

2.1.11 Point 11: Investigating Incidents

Incidents can occur during commissioning and start-up. *Incident investigation* refers to the process of identifying the underlying causes of incidents (including "near misses" that could have had serious consequences) and taking steps to prevent similar occurrences. The point is for employers and employees to learn from experience and thus avoid repeating mistakes. See "Lessons Learned" in Chapter 8.

CSTs who were present during an incident will be interviewed as either witnesses or contributors to the accident or investigation. If the incident is associated with instrumentation or controls, a CST may also be asked where his or her expertise can aid in the investigation. Most employers already have the in-house capability to investigate incidents prior to start-up, and a multidisciplinary team is a good means for analyzing facts and developing plausible scenarios. Team members should be selected on the basis of their training, knowledge, and ability to contribute to a team effort to fully investigate the incident. The investigation should seek to obtain facts, and the process should deal with all involved individuals in a fair, open, and consistent manner.

2.1.12 Point 12: Emergency Preparedness

All project team members should know how to react when an emergency alarm is activated. The employer's training program, normally made part of employee orientation, should address the emergency preparedness training needs of all employees at the plant site. Personnel are not allowed into plant areas without completing this type of comprehensive safety-related training. Better preparedness is accomplished by conducting drills, training exercises, or simulations with community emergency response planners and responder organizations.

As an employee of the company during an emergency, you will be required to follow all evacuation or emergency procedures. As a member of the emergency response team (ERT), you may help attend to the emergency situation and to affected personnel. ERT personnel get additional training in firefighting and lifesaving measures, which are encompassed in the term *emergency response*. OSHA recommends that all employers have an emergency action plan that describes the actions employees should take to ensure their safety in a fire or other emergency.

Relevant Standards

US Occupational Safety and Health Administration

- 29 CFR 1910.38, *Emergency Action Plans*, July 1, 2017

- 29 CFR 1910.120, *Hazardous Waste Operations and Emergency Response*, OSHA, July 1, 2017

2.1.13 Point 13: Compliance Audits

An audit team normally evaluates the company's compliance with the provisions of PSM. Through systematic analysis, the team should document areas that require corrective action as well as areas where the PSM system is working effectively. The CST may be a member of such a team.

2.1.14 Point 14: Trade Secrets

All the information needed to comply with OSHA PSM must be available to those compiling the process safety information; developing the PHA; developing the operating procedures; and performing incident investigations, emergency responses, and compliance audits. Because this may involve potential trade secret conflict of interest, one solution is to require employees to sign confidentiality (nondisclosure) agreements.

2.1.15 Summary: The CST and the 14 Points

Clearly the CST is intimately involved in all aspects of OSHA PSM. Although each plant has specific PSM procedures, most US plants that deal with hazardous materials are basing their procedures on OSHA PSM. For cases where a particular hazard is not addressed by an OSHA standard, the OSHA general duty clause may apply.

OSHA also dictates the topics on which employees must receive annual training. This type of training is known as *regulatory safety training* and will be discussed in Section 2.3.

The following section addresses safety meetings, their content, and who attends and runs them.

Relevant Act

US Occupational Safety and Health Administration

- *Occupational Safety and Health Act of 1970*, Section 5(a)(1), General Duty Clause, also shown as: 29 U.S.C. 654, 5(a)1, 29 U.S.C. 654, 5(a)2, and 29 U.S.C. 654, 5(a)3

2.2 Common Types of Safety Meetings

All manufacturing plants must provide safety training. This requirement is normally accomplished through safety meetings. Many different types of safety meetings and safety training occur in the manufacturing industries, and the content and frequency vary for each. Many plants use their safety meetings to disseminate OSHA PSM information, but other information is passed along as well. The discussion that follows is based, in part, on the author's experience and on an informal survey she conducted to determine the type and content of safety meetings in a representative industry—the chemical industry, in this case.

A certain amount of required regulatory safety training (discussed in Section 2.3) must occur annually. Once these requirements are fulfilled, plants have the leeway to augment their safety programs with other material. They may conduct safety meetings composed of large or small groups of people, or offer one-on-one training (e.g., a safety trainer or engineer with a trainee).

A large group might consist of an entire Maintenance department; a combination of maintenance, operations, and administration personnel or office workers; or a subset of employees in a plant, such as workers producing a particular product or operating a particular process.

A typical small group might be a crew that consistently works a shift together. This type of meeting may occur in the control room before a shift begins or in a change room or lunchroom, and is sometimes moderated by a supervisor or lead operator.

CSTs may meet with small groups of technicians on a consistent or special basis, depending on the work about to be performed.

One chemical plant the author surveyed conducts daily meetings in its I&E shop. Although the shop or area supervisor usually conducts the meetings, sometimes guests, or the workers themselves, may do so.

Small groups working on a special assignment, start-up, or shutdown may meet to discuss the work and associated hazards they may have encountered that day or on the previous shift. One chemical plant in the survey required anyone working in an alkylation unit turnaround to view a video pertaining to hydrofluoric acid safety.

This type of focused, small group meeting has many names including *toolbox, handrail, job briefing,* and *safety stand-down.* These small meetings may also occur at the job site with the employees who are to perform the work, whether it is maintenance or new equipment installation. The intent is to cover the hazards of a particular job prior to starting the work. This is an OSHA requirement, for example, in the power generation industry. This type of meeting is very effective in the start-up environment because it often occurs at the site of the start-up, such as the control room.

"Meetings" consisting of one person to fulfill safety training have become possible with the inception of computer-based training (CBT). With CBT, individuals log on to their computer, take a study course for a particular topic, and may take one or more study quizzes and a final exam. Many plants use CBT because test scores, course attendance, and training frequency can be automatically monitored. CBT also frees up time for group safety meetings on other subjects.

Companies can also ensure that personnel know relevant plant safety information by giving employees access to standard operating procedures (SOPs) and SDSs and testing them individually. This may be managed by self-study or by attending classes. Courses can be conducted off-site for specialized topics such as distributed control systems (DCSs) and instrumentation. However, company information such as the chemicals, safety protocols, and procedures used on-site is normally discussed in a meeting room.

An SOP provides personnel with a step-by-step means to ensure a safe working environment for a specific class of chemicals or type of hazard and to ensure that people perform a job consistently to achieve a high-quality product or high production levels. SOPs may take many forms, such as the one shown in Figure 2-1.

The locations of safety meetings vary by company, and meetings may occur on- or off-site. The location may be determined by the size of the group; the need to minimize interruptions; or the need for access to materials and instructors, consultants, or subject matter experts. The plants the author surveyed indicated that monthly meetings were normally held in a conference room, but short, informal meetings often occurred in a control room prior to the start of a shift. Meetings were also conducted on the plant floor or in a maintenance shop to illustrate specific examples or even safety problems, or to show which SOPs should be followed.

Safety meetings are also held at engineering firms, often in conference rooms, attended by personnel (engineers, managers, technicians, administrative assistants, and purchasing agents) involved with the new or upgraded plant being designed, prior to start-up.

Standard Operating Procedure

Title: Taking a Daily Sample from Tank 203

Section 1 – Process

1.1 Obtain a sample bottle with the proper chemical and safety labeling. Complete all fields on the label (e.g., chemical, source (tank number), date, initials).
1.2 Wearing appropriate PPE, proceed to Tank 203.
1.3 Carefully opening side sample port on Tank 203, allow solution from tank to flow into catch basin for 30 seconds to obtain a well-mixed, representative sample.
1.4 Catch sample, filling the sample jar 2/3 full.
1.5 Close the sample jar tightly.
1.6 Turn on the catch basin pump to allow the sample material to flow back into Tank 203. When the catch basin is empty, turn off pump.
1.7 Bring the sample to the laboratory; and note the collection time in the lab logbook.

Section 2 – Hazardous Chemicals

3% solution acetic acid

Section 3 – Approvals Required

Signature: _____ Date Approved: _____
Signature: _____ Date Approved: _____

Section 4 – Designated Area

Area 200

Section 5 – Special Handling Procedures and Storage Requirements

5.1 Label all sample materials with their full name (i.e., acetic acid solution, not AA).
5.2 Store only empty clean and unused sample bottles in the samples storage cabinet. Use only clean sample bottles from this source to obtain sample(s).
5.3 Use secondary containment carriers whenever transporting hazardous material outside of the lab. Use due care and caution when moving hazardous materials around anywhere.

Section 6 – Personal Protective Equipment

6.1 Eye Protection
- Safety glasses or goggles
- Face shield if desired

6.2 Protective Clothing
- Apron or lab coat
- Gloves: nitrile, butyl, PVC or relevant material

Section 7 – Engineering/Ventilation Controls

Use a fume hood in the laboratory. Perform all operations in the hood, stand behind the sliding windows and reach around to perform the manipulations required. Dispose of unused sample per Procedure Lab203.

Section 8 – Spill and Accident Procedures

In the case of a spill:
8.1 Avoid breathing vapors.
8.2 Quickly identify the spilled material if you can do so safely.
8.3 Alert people in the area, use caution tape and closed doors to keep people out of the spill area, where applicable.

Section 9 – Waste Disposal

Dispose of the properly labeled waste in a safe and legal manner. Nonhazardous waste may be placed in a container to go to a sanitary landfill or, if appropriate, washed into the sewerage system.

Section 10 - Decontamination

Properly decontaminate the area by cleaning up most of the loose material after a neutralizing agent (higher pH such as soda ash/sodium bicarbonate). Kitty litter can allow excess material to be shoveled up.

Figure 2-1. Generic SOP form.

Typically, safety meetings occur either early in the morning or late in the afternoon to accommodate both shift and office workers. Hourly personnel must sometimes "work over" from the previous night for the meeting, clocking out after the meeting is over. The same applies for those "working over" from the afternoon shift. A CST may be working a rotating shift, straight days or nights, or even be on vacation, but attendance at these meetings may still be mandatory. According to the survey, companies may not be required to pay employees for the extra time spent attending these training meetings, but they often do.

The frequency of safety meetings may vary among organizations, but the amount of regulatory safety training that is required generally necessitates that these meetings be held at least monthly. Additional individual CBT and off-site training often occur during the year as well.

Documenting these meetings is important for the company and the employees. The documentation proves that employees have been told how to do their jobs safely, which is accomplished through OSHA-required training as well as site-specific training. Sign-in sheets, password-protected logins for CBT, and even diplomas or certificates offer proof that plant personnel have attended the appropriate safety meetings and courses. All the plants in the informal survey noted that attendance at their training meetings was fully documented through the use of sign-in sheets. The minutes of the meetings were also recorded and filed, and tests taken electronically were maintained and checked by the plants' Training departments. In addition, these meetings allow employees the opportunity to relay safety and health concerns and improvements as well as business policy information to their peers and supervisors.

2.3 Common Types of Safety Training

As discussed earlier in this chapter, OSHA mandates that a certain amount of regulatory safety training occurs annually, much of which is also relevant to start-ups. A plant may also wish to augment its safety program with additional safety training specific to the plant and its employees' personal lives.

2.3.1 OSHA Regulatory Safety Training

OSHA regulatory safety training topics are as follows:

- Temporary barricades
- Employee exposure and medical records
- Basic electrical safety
- Fire extinguisher basics
- Hazardous waste operations and emergency response (HAZWOPER) awareness
- Personal protective equipment (PPE)

- Ladders, stairways, and scaffolding
- OSHA hazard communication standard
- Blood-borne pathogens
- Hearing protection
- PSM
- Confined-space entry
- Respiratory protection
- SDSs
- LOTO, hot work, and line-breaking

Of these, the topics that are particularly relevant to the CST are basic electrical safety, PPE, confined-space entry, SDSs, LOTO, hot work, line-breaking, and ladders, stairways, and scaffolding. OSHA requires that each of the listed topics be taught once per year. Following is a brief summary of the covered subjects.

2.3.2 Industry-or Plant-Specific Training

The demands of each plant will dictate the frequency and content of this type of training. Each topic is typically addressed once per year, or more often if problems are occurring. The topics may be discussed along with other safety training material and topics, regulatory or not; therefore, more than one topic may be on the meeting agenda. Examples of the training specific to a particular industry or plant are as follows:

- Recent incidents, near misses, lost-time accidents, and first aid
- New and existing policies and procedures
- Alerts on health issues from the plant nurse or doctor
- Unsafe working conditions
- How to deal with a chemical specific to the site
- Specific company or job-related topics

For special equipment and PASs, hands-on training enhances learning. New personnel, for instance, benefit by being trained on an operator training simulator. Upset conditions of various types can be displayed on the simulator, and the trainee can go through the proper operating procedures to bring the system, which represents the plant, back to normal. This realistic training can be effective in teaching employees. Through this type of training, operations personnel gain an understanding of the process and how the instruments and controls function for start-up and normal maintenance.

Other training techniques using videos or on-the-job (rather than simulation) training can also be very effective for teaching job tasks, duties, and other important tasks because they are site-specific. An effective program will allow employees to fully participate in the training process, to develop and practice their skills, and to use their new knowledge.

Some examples of plant and/or job-related ISA standards that may pertain to the plant you will work in are cited below.

Relevant Standards and Recommended Practices

American National Standards Institute/International Society of Automation

- ANSI/ISA-60079-0 (12.00.01)-2013 (R2017), *Explosive Atmospheres – Part 0: Equipment – General Requirements*
- ANSI/ISA-12.01.01-2013, *Definitions and Information Pertaining to Electrical Equipment in Hazardous (Classified) Locations*
- ANSI/ISA-12.04.04-2012, *Pressurized Enclosures*

International Society of Automation

- ISA-12.10-1988, *Area Classification in Hazardous (Classified) Dust Locations*
- ISA-RP67.04.02-2010, *Methodologies for the Determination of Setpoints for Nuclear Safety-Related Instrumentation*

American National Standards Institute/UL, LLC

- ANSI/UL 12270-2017, *Requirements for Process Sealing Between Electrical Systems and Flammable or Combustible Process Fluids*

2.3.3 Safety Programs Evaluation

Employers must periodically evaluate their safety training programs to determine whether the necessary skills, knowledge, and routines are being understood and properly implemented by their trained employees.

Careful consideration must be given to ensuring that employees—including maintenance and contract employees—receive current and updated training. For example, if a change is made to a process, the employees affected by the change must be trained to understand it and its effects on job tasks (e.g., new operating procedures). Again, evaluating the training's effectiveness is crucial.

Leadership

People who conduct safety training may be consultants, safety engineers, trained plant personnel, area supervisors, safety team leaders, or specialists. Audiovisual aids such as videotapes, slides, and computer presentations may accompany a lecture or interactive discussion. Simulations of events, such as a fire drill or emergency response, may also be a part of a plant's safety program. All of these formats are typically used randomly throughout the year to make a plant's safety program interesting and informative.

Volunteer personnel from the plant may be members of an ERT, leaders of a safety team, or safety auditors.

Safety team leaders may be volunteers in the plant organization, degreed safety engineers, or people whose position in the company (safety coordinator or trainer) includes this duty. If an employee sees anything unsafe, or if parts of the job entail unacceptable risk, the employee should discuss it with the offending party, a supervisor, or other appropriate personnel.

JSA

Before a job is performed, many companies require that a JSA be performed. A JSA is a structured, step-by-step analysis of the steps taken to do a particular job, with an emphasis on the safety of each step or job task. Many of the tasks a CST does during the commissioning and start-up phases of the project may require that a JSA be performed using the companies' standard procedures and forms.

2.4 LOTO

Of the many safety procedures associated with commissioning and start-up, one of the most important that the CST will be involved in is LOTO. LOTO prevents accidents associated with energy sources such as electrical, pneumatic, hydraulic, steam, and gravity-fall.

Lockout normally means that physical locks are placed on all energy sources for the equipment to be worked on. It can also mean that wires are disconnected or items are tagged out of service. Often a permit must be filled out for the equipment to be shut down. For example, a lock may be placed on a starter in the motor control center to prevent machinery from being started. All people working on locked-out equipment must use their own personal lock and key.

Tagout normally means that a tag stating how long the lock is to remain in place is attached to the lock. The tag is signed and dated. The CST may have to lockout a piece of equipment he or she is working on, or may need to be aware that the equipment has been locked out. The permit and tag are valid only for a fixed period of time, most likely an 8- or 12-hour shift.

LOTO precautionary measures can be simple or complex. Examples include the following: a permit is required; lines must be emptied or vessels purged; sniffers, gas detectors, or detectors specific to a type of chemical are used to detect whether any hazardous conditions exist; the power source must be de-energized, and equipment tested to make sure it will not start; a "fire watch" must attend in case a fire extinguisher is needed.

Hardwired (hardware) or soft wired (software) interlocks may be involved, so a complete LOTO procedure ensures that these interlocks will not contribute to an unsafe work situation. In the case of safety during loop checking, the CST must know whether hardware or software interlocks are in place and how they are controlled. These should be shown in detail on P&IDs, loop sheets, and electrical drawings. An understanding of how these loops are controlled enables those involved with the LOTO procedure to de-energize and then work on the equipment safely.

Example sequence of events during LOTO:

1. Shut down the operation.
2. De-energize electrically run equipment.
3. Install lock(s) on applicable motor starter(s).
4. Test for power before starting work.
5. Block and bleed lines/process.
6. Drain lines and install blanks/blinds.

Before work begins on a locked-out piece of equipment, connections to *all* energy sources must be broken, any stored energy must be released, and the equipment must

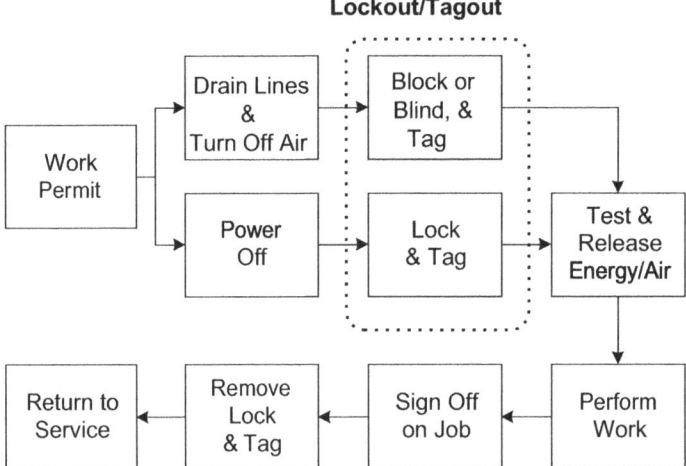

Figure 2-2. LOTO block flow diagram.

be tested to make sure it is inoperative. As a CST, you must be trained in these procedures, and you must understand and practice them.

Figure 2-2 depicts LOTO procedures for electrical and nonelectrical (air-powered) equipment.

Relevant Standard

<u>US Occupational Safety and Health Administration</u>

- 29 CFR 1910.147, *The Control of Hazardous Energy (Lockout/Tagout)*, July 1, 2017

2.5 Compliance Documentation

Every plant is different, so this discussion of compliance documentation will be general. However, the plants informally surveyed for this book provided the following specific examples of documentation:

- **Safety** – Documents such as SDSs and container labels
- **Policy** – A company's safety and/or environmental pledge(s) as stated on company letterhead by the company chief executive officer
- **Procedural** – SOPs and LOTO procedures, etc.
- **Environmental** – Permits and emission monitoring

Safety documentation may be hard copy or in electronic form. Here we are particularly concerned with safety procedures and rules, operating and maintenance procedures, and SDSs.

OSHA requires that safety-related documents be accessible to all shifts with no barriers to accessibility.

Because CSTs must understand and follow safety procedures, they must comply with compliance documentation requirements. For example, the CST may need to assist the environmental engineer who prepares the toxic chemical release inventory for the yearly Superfund Amendments and Reauthorization Act (SARA) Section 313 requirements.

Relevant Standard

US Occupational Safety and Health Administration
- 40 CFR 372, *Toxic Chemical Release Reporting: Community Right-to-Know*, July 1, 2017

2.5.1 SDSs

An SDS is a document pertaining to a single hazardous chemical. National and international requirements for SDSs vary. In the United States, SDSs are created and distributed in accordance with OSHA CFR 1910.1200 (g). There must be an SDS for *each* hazardous chemical in the plant.

Although there is no standard format for SDSs and they vary with the manufacturer or distributor of each chemical, OSHA regulations list the types of information they must contain.

Because equipment may come in contact with hazardous chemicals during startup, it must be handled and operated appropriately. For example, if a CST has removed an instrument from service, it may need to be handled with gloves to be cleaned, or have parts replaced with appropriate materials so it may be used with a given chemical.

A piece of equipment (or an instrument) that is being returned to a vendor must have a signed document stating the service the equipment was used in and what materials or chemicals the equipment has been in contact with. Preparing this document may be part of the CST's job. This is one reason that it is important for you to know how to read SDSs, where they are located, and which chemicals might be encountered in a particular plant area and in particular equipment. The plant provides this information as part of the OSHA hazard communication standard, but employees, including CSTs, must take responsibility for reviewing the information themselves.

An example SDS is shown in Appendix B.

2.6 Frequently Encountered Safety Equipment

The appropriate safety equipment depends on the type of plant it is used in. Most plants require basic PPE such as a hard hat, safety glasses, and steel-toe or leather shoes, but other requirements may vary. Plants with dust- or vapor-laden atmospheres, or the potential for them, may require that some type of respiratory protection be worn on the face all the time, worn around the neck and used as necessary, or kept handy in a bag in case of emergency. Special (explosion-proof) tools and equipment for these types of atmospheres are also required to prevent an explosion. There should be a plant policy for the use and availability of PPE, tools, and equipment, as well as for their upkeep.

Some plants may require the use of fire-retardant clothing (FRC), such as Nomex, or coveralls. For example, the power and pipeline industries encourage and often require employees to wear cotton clothing. Natural gas pipeline companies, chemical companies, and refineries require Nomex FRC in process areas.

Examples of other PPE that might not be used all the time include "acid suits" for unloading acids or bases, gloves, face shields, and hearing protection (earplugs and earmuffs).

Plant safety equipment includes fire extinguishers and fire suppression systems; explosive and flammable gas detectors, also known as *sniffers*; harnesses; fire blankets; safety showers; fire monitors; deluge systems; self-contained breathing apparatuses (SCBAs); and rescue stretchers.

Halon or other (post-1994) air-displacing fire suppression systems may be present in electronic or computer-based control rooms or other electrical equipment rooms where fire suppression by water is not applicable. (Halon production ceased in 1994 in most countries, including the United States, due to chlorofluorocarbon-linked ozone layer depletion concerns.) Because PAS equipment is normally protected by these types of systems, the CST would be one of the people required to evacuate.

Fire and gas detection systems are also standard systems that not only detect and alarm problems but also are often connected to the main PAS. Therefore, the CST may be involved with this interface.

Harnesses are typically found in the maintenance supply area and are used during jobs that require fall protection or for tank entry procedures. If a CST has to work on elevated surfaces, he or she may have to wear a harness and be trained in its use.

2.7 Start-Up Safety: Practical Examples

The following sections describe typical start-up scenarios that involve safety. Would you be ready to deal with them?

2.7.1 PHA

A new tank is being installed to handle additional sulfuric acid inventory in the plant. Figure 2-3 depicts a tank with a capacity of 5000 gal (18,926.5 L) and with automatic valves on the fill and discharge lines, an overflow line, and a level indicator. There is also an interlock between the high-level switch and the fill valve. Questions that might come up during a PHA include the following:

- Can an entire tank truckload fit into this tank?

- If the tank is not empty and a truck is to be unloaded into this tank, how low must the level be to fit the entire truckload in the tank?

- What is the level switch high (LSH) set for?

- Does a local level indication exist?

- How does the person unloading a truck know how much is in the tank?

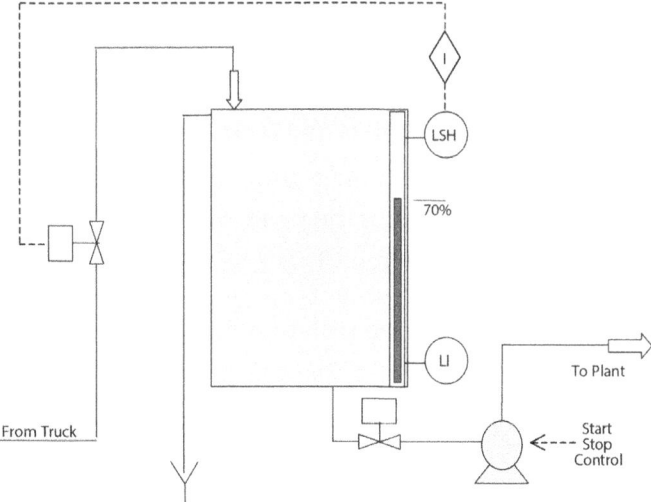

Figure 2-3. 5000-gallon (18,926.5-liter) tank.

- Can another chemical be accidentally unloaded into this tank? What can be done to prevent this?

- What chemicals can mix with this chemical if the tank should overflow to the ground?

- Is the tank in a diked area?

- Is there soda ash or some other neutralizing agent in the overflow area?

- Are acid flanges installed throughout the system?

2.7.2 Emergency Drill or Evacuation

Suppose the wrong chemical was unloaded into a sulfuric acid tank. It was hydrochloric acid. A cloud of fumes is coming out of the tank. What happens next?

- Personnel in the area who first see the problem immediately call their supervisor or safety coordinator.

- Operations is notified (and may shut down the affected area).

- Emergency response procedures commence.

- A shelter-in-place alert may be announced via the emergency system. The local emergency response technician may take action to handle the situation, or the local authorities may be summoned to assist.

- Plant management calls the US EPA and local radio stations as necessary.

- Nonessential personnel may be told to go home.

- After the immediate danger has passed, an investigation is conducted by appropriate plant personnel.

2.7.3 Deciding When a Respirator Is Necessary

Suppose you have a work order to repair a transmitter in the plant. It is important that this work be done now. On entering the plant, you notice a lot of dust in the room atmosphere. How should you proceed?

- First, notify Operations about the situation. If you do not have a respirator, you should not enter this area until the dusting condition is remedied.

- If you have a respirator and it has been fit-tested, then you should check that it fits properly (plants have requirements about being clean shaven to wear a respirator properly), put it on, and after completing the job, clean and store it properly or dispose of it (if it is a disposable type).

2.7.4 Proper LOTO Procedures

A powered piece of equipment must be locked out before being worked on. How is this done? As described in Section 2.4, the following steps are typical:

1. A work permit and tag should be filled out properly, dated, and signed.

2. Every possible energy source should be turned off and any captured energy, such as air pressure or gravity drop, should be released.

3. A lock should be applied to the switch, valve, or other device, that controls each energy source by, at minimum, the person who would be in danger if there was an accidental start. Depending on the plant rules, multiple personnel may apply their own locks.

4. The lockout should be tested determine if the equipment will start.

5. After work is complete, the lock can be removed by the same person(s) who put the lock(s) on, and the sources of energy restored.

2.7.5 PSSR

A PSSR is a systematic review of a process prior to the introduction of highly hazardous chemicals and subsequent start-up of the process. This topic is discussed later in this book.

2.8 Safety Instrumented Systems

One component of the overall PAS is the SIS. The SIS performs specified functions or safety instrumented functions (SIFs) as part of an overall risk reduction strategy. This topic will be discussed more fully later in this book.

Summary

There are many rules that companies and their employees must follow to comply with OSHA PSM and other requirements, as well as those of other agencies and organizations. These rules are disseminated through training and apply during start-up and all other times the plant is in business, including during a shutdown or turnaround.

Many changes occur during start-up, and new information and company policies are constantly being made available. The lines of communication must be open during this critical and potentially dangerous phase of a plant's life. If information is not forthcoming then you, the CST, must ask questions and seek the answers. If you feel that safety is being compromised, then you must pursue the matter with managers and those in the Operations department. Your CST skills will help get the new plant online, but you must also acquire safety skills through training and safety awareness activities to ensure personal and plant well-being.

Review

2.1 A calciner with a local burner management system (BMS) has a local panel controlled by a programmable logic controller (PLC). The flame does not stay lit. Someone is suggesting that the purge time is too long and that the pilot is being blown out. He suggests testing by "jumpering out" an interlock to prove this. What does he mean by this? Why is this wrong? How can you fix this problem? What documents might you use to work on this job?

2.2 You have been told you must attend a process hazard analysis (PHA) meeting about a change occurring in Area 3 of the plant. What should you do to prepare for the meeting? What might you bring to the meeting? Why were you invited to this meeting?

2.3 It is 3:30 p.m. on Friday. The plant operations manager has decided that he wants a 150 hp (111,855 W) blower replaced with a 300 hp (223,710 W) blower because not enough airflow is being provided to the fluid bed dryer. He thinks the job should not take more than about 4 hours, and he wants production to be back online by 11 p.m. What are some of the factors that might compromise the safety of this job? What should be done to ensure that maintenance personnel can work on the blowers safely? What might you as a control CST be involved with in connection with this task?

2.4 What government organization created the *Process Safety Management of Highly Hazardous Chemicals* standard?

2.5 What is the objective of process safety management (PSM)?

2.6 Name two engineering societies that provide technical reports that help maintain good engineering practice.

2.7 What document gives information for determining which chemicals are hazardous and how to deal with them safely or eliminate them completely?

2.8 Name some drawings that are important during a PHA and pre-startup safety review (PSSR).

2.9 Is following Management of Change (MOC) procedures required when replacing a pump with another one of the same type?

2.10   Why is it important to consider materials of construction when working on a job or replacing parts?

2.11   What is another term for *nonroutine work*?

2.12   What is the sequence of events necessary for the proper lockout of a pipeline attached to a pump possibly filled with a (liquid) chemical?

2.13   Name two important documents used during lockout.

2.14   Identify some documents that would need to be changed as a result of a process change.

2.15   Name some teams (including volunteer organizations) a CST may be involved with.

2.16   Define a *near miss*.

2.17   Name some types of safety training that plants offer.

2.18   What training is mandatory? Name some of the topics covered by such training.

2.19   Why can start-ups be particularly hazardous?

2.20   Name some safety documents.

2.21   Name five pieces of PPE.

Recommended Reading

Books

Bryan, Austin, Elizabeth Smith, and Kevin Mithcell. *Performance-based Fire and Gas Systems Engineering Handbook*. Research Triangle Park, NC: ISA (International Society of Automation), 2016.

Goble, William M., and Iwan van Beurden. *Safety Instrumented Systems Design: Techniques and Design and Design Verification*. Research Triangle Park, NC: ISA (International Society of Automation), 2018.

Gruhn, Paul. "Safety Instrumented Systems in the Process Industries." In *A Guide to the Automation Body of Knowledge*, 3rd ed., ed. Nicholas P. Sands and Ian Verhappen. Research Triangle Park, NC: ISA (International Society of Automation), 2018.

Gruhn, Paul, and Simon Lucchini. *Safety Instrumented Systems: A Life-Cycle Approach* Research Triangle Park, NC: ISA (International Society of Automation), 2019.

Magison, Ernest. *Elecrical Instruments in Hazardous Locations.* 4th ed. Research Triangle Park, NC: ISA (International Society of Automation), 1998.

Magison, Ernest, updated by Ian Verhappen. "Safe Use and Application of Electrical Apparatus." In *A Guide to the Automation Body of Knowledge*, 3rd ed., ed. Nicholas P. Sands and Ian Verhappen. Research Triangle Park, NC: ISA (International Society of Automation), 2018.

Marszal, Edward M., and Eric W. Scharpf. *Safety Integrity Level Selection: Systematic Methods Including Layer of Protection Analysis.* Research Triangle Park, NC: ISA (International Society of Automation), 2002.

Standards

ANSI/ISA-84.91.01-2012. *Identification and Mechanical Integrity of Safety Controls, Alarms, and Interlocks in the Process Industry.* Research Triangle Park, NC: ISA (International Society of Automation).

ISA-TR84.00.04-2015. *Part 1, Guidelines for the Implementation of ANSI/ISA-84.00.01-2004 (IEC 61511).* Research Triangle Park, NC: ISA (International Society of Automation).

ISA-TR84.00.04-2005. *Part 2, Example Implementation of ANSI/ISA-84.00.01-2004 (IEC 61511 Mod).* Research Triangle Park, NC: ISA (International Society of Automation).

ISA-TR84.00.07-2018. *Guidance on the Evaluation of Fire, Combustible Gas, and Toxic Gas System Effectiveness.* Research Triangle Park, NC: ISA (International Society of Automation).

ISA-TR91.00.02-2003. *Criticality Classification Guideline for Instrumentation.* Research Triangle Park, NC: ISA (International Society of Automation).

3
Documenting the Commissioning and Start-Up Processes

Project Documents

Hard Copy and Electronic Documents

Document Locations

Commissioning and Start-Up Drawings and Documents

Documents Used by the CST

Control of Project Documents

This chapter will describe the documents[1] typically available and necessary to gain an understanding of the scope of the CST's job during a project and subsequent start-up, and any other time the CST is involved in plant activities. The types of documents, and their location, use, and control will be described. The chapter will also look at various formats in which documents may be available, including electronic and hard copy.

Many of these documents are needed during the project and start-up and sometimes must be updated as start-up proceeds. The CST must know how to mark up

---

1 For clarity, wherever possible throughout this chapter, we will refer to drawings as *drawings* and to all other documents as *documents*. Where is it clear from context, and to minimize the repetition of *drawings* and *documents*, *documents* may include drawings.

these documents and whom to inform when they need updating and reprinting. A company might not require that CSTs use all these documents, but by reviewing nonrequired documentation during idle times CSTs will enhance their knowledge of the process and understanding of their place in the overall plant "picture."

It is vital to document and thereby verify that commissioning and start-up activities have been completed. Following a good factory acceptance testing (FAT) plan, accurately documenting test results, and documenting that calibration and loop checking have been completed help ensure minimal rework and an efficient and successful start-up.

Without documentation, verification, and sign-off, calibrations may have to be redone, loops may not perform properly, paperwork may not be available when needed, and time may be used inefficiently, resulting in costly delays.

There are several ways to manage documentation pertaining to calibration and loop and function checking, and methods can even vary among plants within one manufacturing facility—especially if a plant standard has not been established and the facility is not regulated by FDA or ISO requirements. The information in this chapter is based on the author's personal experience in manufacturing facilities and a survey of the ways other chemical companies completed calibration and loop checking prior to starting up their plants.

Calibration and loop and function checking are, of course, complex subjects unto themselves and are described in detail in the ISA Technician Series by Mike Cable and Harley Jeffery (see the "Recommended Reading" section in this chapter).

3.1 Project Documents

Many types of industrial information must be documented, which results in the preparation of drawings, tables, procedures, charts, and descriptive text that describe how the plant was built, how it is run, and how to maintain it safely and efficiently. Usually, these documents adhere to standardized formats, which include a template or form for writing procedures and industry standards for making drawings. It is important to abide by these standardized formats for two reasons. First, people have been trained to read documents a certain way, and using the standard format helps ensure they will understand and employ the documentation properly. Second, it is easier to change and update the documents if consistency is maintained. These documents will be discussed in further detail later in this chapter.

3.2 Hard Copy and Electronic Documents

Some companies have an extensive plant information network (PIN) or computer network (also referred to as a *business network*) to which drawings and other documents are added as they are updated. Some companies still have and use printed documents, although to a lesser extent today. Personnel working in a department commonly known as *Document Control* usually maintain all of these documents.

A CST must know how to get the latest copies of these documents and how to use the tools to maintain them, such as computers, copy machines, and files. A CST must also understand why there are several copies of what seem to be the same document and how to distinguish them.

3.2.1 Original (Hard-Copy) Drawings

In plants that do not have computer-based documentation, people may look at original drawings, such as piping and instrumentation drawings (P&IDs), and make copies for their own use, or they may distribute copies to various departments where they are kept in specified locations. P&IDs of this sort are usually very large. To make them easier to handle, especially in the field, they are commonly reduced to 8.5 in × 11 in (letter size, approximately size A4) or 11 in × 17 in (size B) using a copy machine. Additional sizes include C (22 in × 17 in), D (22 in × 34 in), and others. The text and figures should be legible when the P&ID is reduced. You may find that smaller drawings are easier to work with during start-up and troubleshooting because you can carry them to where you are working.

In the past, original drawings such as P&IDs were produced by draftspersons on vellum sheets and kept in large drawers in a secure location. Many of these vellum drawings are still being used, especially in older plants. Vellum sheets are often fragile, and a company may require that specific people be responsible for their care and use.

Over the years, original drawings are typically updated and copied, and dates are entered that reflect which changes were made and by whom for P&ID revision information (see Figure 3-1 left side).

If these drawings are up to date, a company may opt to scan them into the PIN and allow people to view them there. After a drawing is scanned, a qualified individual must review the original drawing to determine that the scanning process picked up everything.

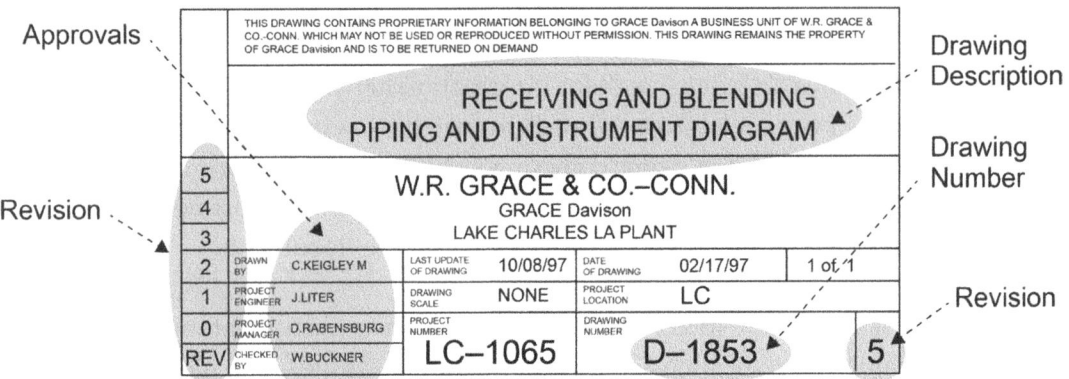

Figure 3-1. P&ID title block.
Source: Courtesy of Grace Davison, Lake Charles, LA.

Though several software packages are available for scanning, there are some drawbacks to using this method. First, these drawings are very large and thus often not compatible with smaller scanners. In addition, scanned drawings cannot be updated as easily as the originals can be when plant changes occur. Also, some scanned documents that have been added to the network and must be accessible to the start-up team may not have searchable text (document content) capability. This depends on whether the scanning method provides optical character recognition (OCR) and has properties compatible with the electronic document management system (EDMS).

Other items (highlighted gray) in Figure 3-1 include the following:

- **Drawing description** – Normally, this is a description of the plant area that the drawing depicts.
- **Drawing number** – This is a unique drawing number for the plant area being depicted. The document can be located by searching for this number in the project drawing database in which the drawing is logged.
- **Revision** – This indicates the version of the drawing. It is important to know the current revision number to ensure that you are working with the latest drawing.
- **Approvals** – The initials of people who are designated drawing approvers indicate that the drawing is approved for use by the project.

3.2.2 Electronic Drawings

Most often, plant personnel who want start-up drawings to be available electronically have them drawn (or redrawn, if existing) using computer-aided design (CAD) software packages. CAD software requires computers with fairly large storage and memory capacities, as well as specialized keyboards. CAD operators, sometimes titled *drafters* or *designers* who may be company employees or contractors, must often go into the field to verify their work when redoing a drawing—especially a P&ID—and it takes a long time to complete such conversions.

CAD standards improve productivity and enable the interchange of CAD drawings between different offices and CAD programs, especially in architecture and engineering firms. Typically, a plant (or corporate) standard is established that determines how these drawings should be created. This involves setting up a legend sheet showing typical equipment, instrumentation, wiring, interlocks and other logic, and equipment information. The standard also dictates how these objects are to appear on the drawings. Examples of wiring and other connections depicted on P&IDs include pneumatic, electrical, hydraulic, electromagnetic, and logical. Standard symbology, based on ANSI/ISA-5.1-2009, *Instrumentation Symbols and Identification*, is normally found on the first drawing in a set of P&IDs, the *legend sheet*, also known as the *lead sheet* (Figure 3-2). Table 3-1 shows in more detail information presented in the Instrument Line Symbols section in Figure 3-2.

After a drawing standard is established, typical drawing elements may be saved as a library of objects, making them easy to use repeatedly in other drawings. Contractors who do CAD work operate under the company's control, and consistent standards make it easier for plant personnel to understand various documents and to train people to use them. Plants can also adopt existing architecture, engineering, and construction standards.

Relevant Standards

British Standards Institution

- BS 1192:2007 + A1:2015, *Collaborative Production of Architectural, Engineering, and Construction Information, Code of Practice*

International Standards Organization

- ISO 13567-1:2017, *Organization and Naming of Layers for CAD – Part 1: Overview and Principles*
- ISO 13567-2:2017, *Organization and Naming of Layers for CAD – Part 2: Concepts, Format, and Codes Used in Construction Documentation*

National CAD Standard Project Committee

- US National CAD Standard, V6, 2014, https://www.nationalcadstandard.org/ncs6/ordering.php

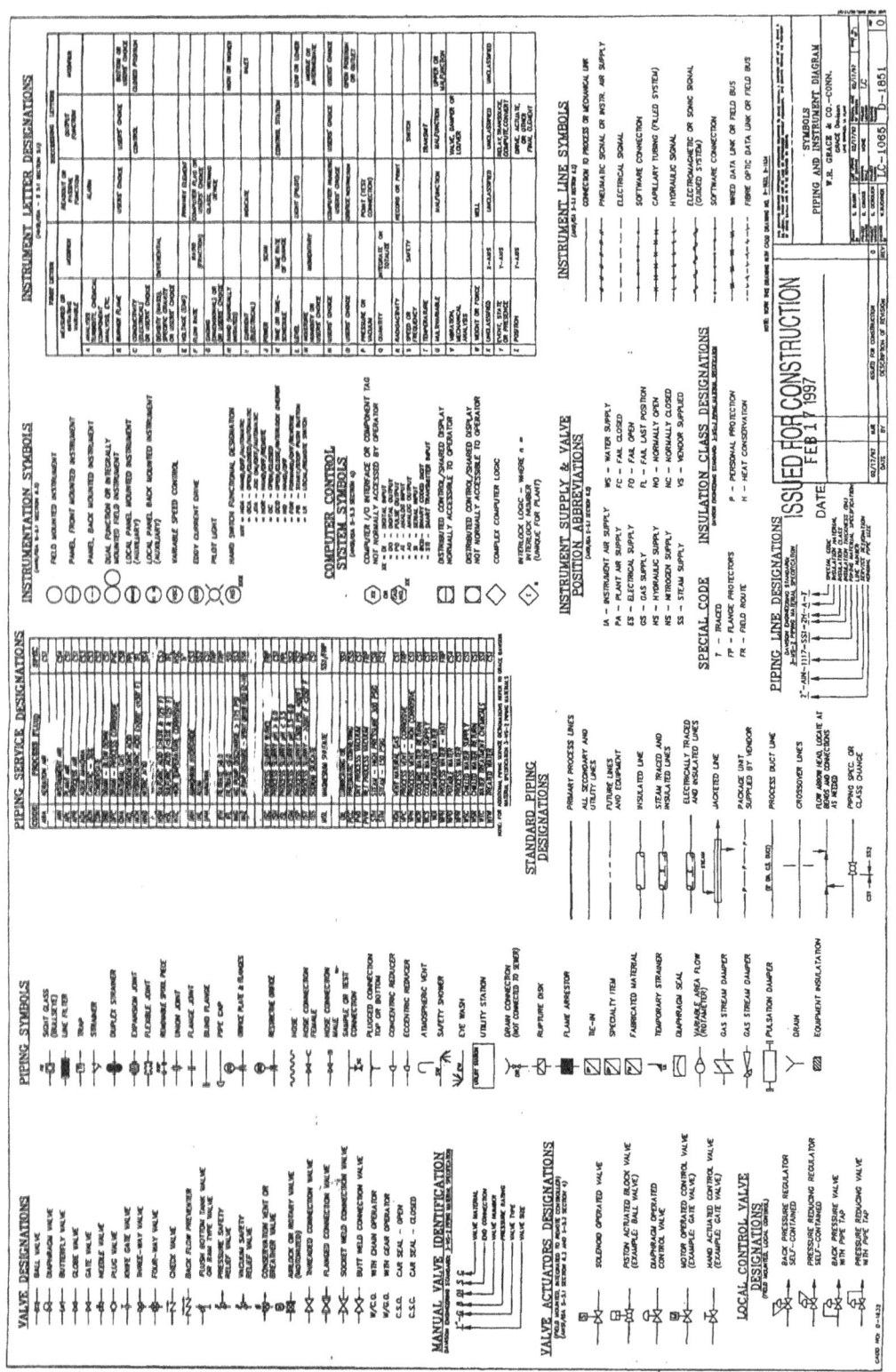

Figure 3-2. Legend sheet.

Source: Courtesy of Grace Davison, Lake Charles, LA.

Table 3-1. Line symbols: Instrument-to-instrument connections.

Symbol	Application
—/—/—	• Undefined signal. • Use for process flow diagrams. • Use for discussions or diagrams where type of signal is not of concern.
—//—//—	• Pneumatic signal, continuously variable or binary.
— — — — — —	• Electronic or electrical continuously variable or binary signal. • Functional diagram binary signal.
———————	• Functional diagram continuously variable signal. • Electrical schematic ladder diagram signal and power rails.
—L—L—	• Hydraulic signal.
—✕—✕—	• Filled thermal element capillary tube. • Filled sensing line between pressure seal and instrument.
—∿—∿—	• Guided electromagnetic signal. • Guided sonic signal. • Fiber optic cable.
a) ∿ ∿ b) ∿⃗ ∿⃗	• Unguided electromagnetic signals, light, radiation, radio, sound, wireless, etc. • Wireless instrumentation signal. • Wireless communication link.
—o——o—	• Communication link and system bus, between devices and functions of a shared display, shared control system. • DCS, PLC, or PC communication link and system bus.
—•——•—	• Communication link or bus connecting two or more independent microprocessor or computer-based systems. • DCS-to-DCS, DCS-to-PLC, PLC-to-PC, DCS-to-Fieldbus, etc. connections.
—◇——◇—	• Communication link and system bus, between devices and functions of a fieldbus system. • Link from and to "intelligent" devices.
—-o—— —o— —	• Communication link between a device and a remote calibration adjustment device or system. • Link from and to "smart" devices.

Source: Reproduced with permission from ANSI/ISA-5.1-2009.

Typical standards for P&IDs include:

- Number of layers (electronic) on the drawing

- Lines where the thickness indicates how process flows are to be depicted

- Scale

- File names, text, and notations

- Revisions and annotations
- Symbols
- Drawing references

Other drawings have many of these same features. Exceptions include electrical drawings that do not include process instrumentation and process flow streams.

Layers

A CAD drawing may have different *layers*, which show different sets of information. This makes it easy to maintain or to print the drawing in various formats. To show temporary changes and changes in progress (revisions), plants use bubbles or clouds (on these drawings) that can easily be removed when the work is completed. See Figure 3-3, which contains a cloud around changes for the last revision of the document. Layers are also convenient when you want to print copies that do not contain all the information because a complete drawing is not needed. For instance, one person may not need to see all the instrumentation, while another may not need to see all the equipment information at the bottom of a P&ID. The plant standard dictates the way in which the information is divided among the layers.

Lines

Major process lines may be drawn heavier or thicker than minor ones. This makes it easier to see the different process flows, especially on P&IDs that are in black and white and contain many lines. For example, a drawing depicting a distillation column might show the major process flows to and from the column (feed, overheads, and draw-off) as heavier lines than the steam and cooling water to the reboiler and condenser, respectively. Process flows are usually shown entering the drawing from the left and exiting at the right, space permitting. In addition, the legend (see Figure 3-2) indicates the symbology used for different line types (e.g., pneumatic, electrical, process flow, and logic).

Scale

Any drawing that depicts physical objects that are larger than the paper size must be scaled to fit the page. A location drawing or plot plan will be drawn to scale. P&IDs, however, are often not drawn to scale. Diagrams, such as flowcharts that do not depict physical objects, are unscaled. Many electrical drawings do not require a scale, but panel layouts are an exception. An example of a drawing scale is 1/4 in = 1 ft (1:48).

File Names, Text, and Notations

Every company has a standard for how documents are named and numbered. An EDMS aids in this effort. Good file maintenance ensures that change tracking occurs

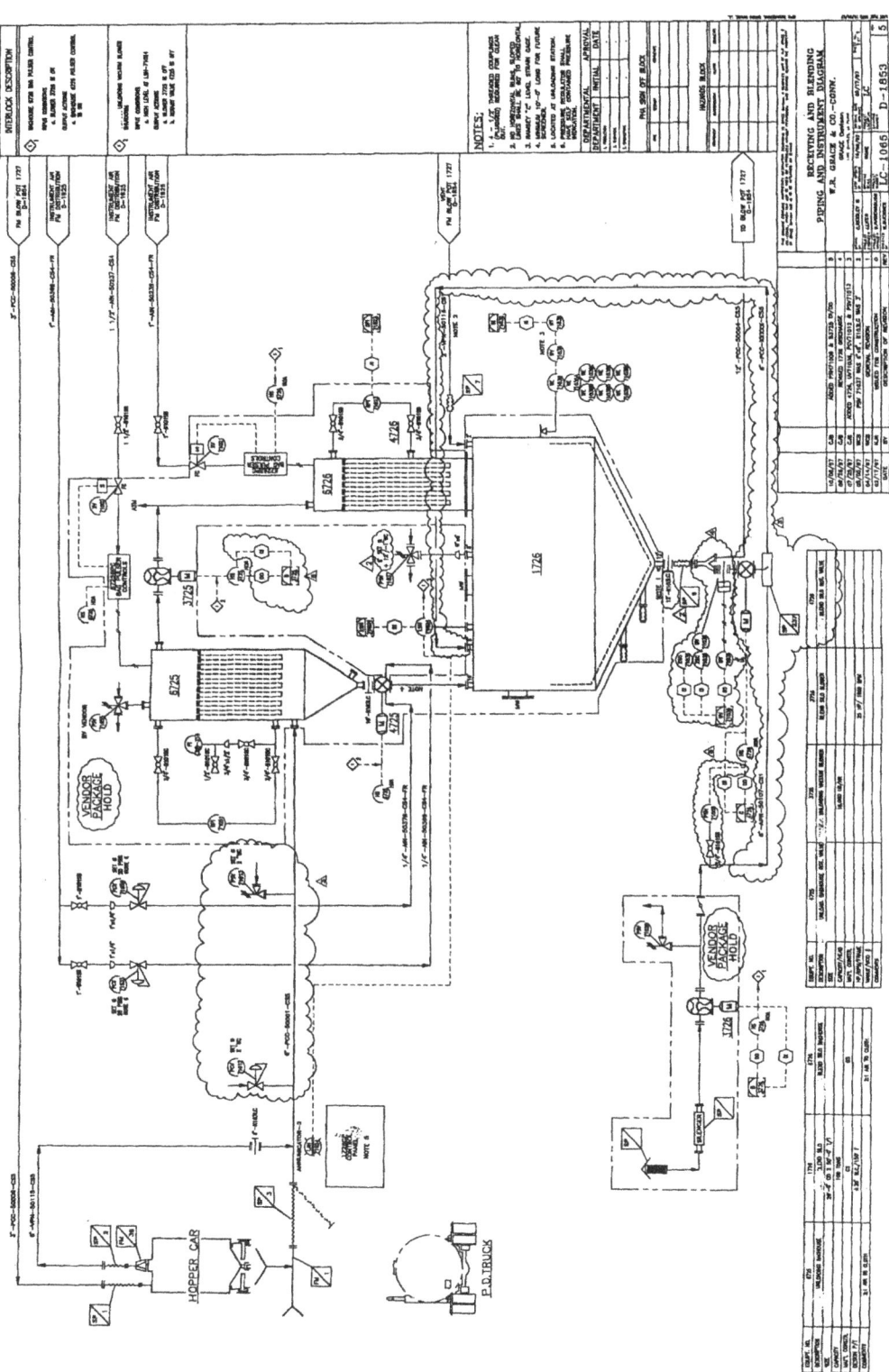

Figure 3-3. P&ID illustrating clouds.

Source: Courtesy of Grace Davison, Lake Charles, LA.

(Management of Change—MOC), standards are adhered to, and documents and drawings are easy to locate. If, for example, two documents had the same number, that would be very confusing. The file name, including version, date, and, where applicable, the names of those who made the change(s) and reviewed or approved the document should appear on the document. Additional text and notations include clouding with symbology to reflect changes from a previous drawing version.

Revisions and Annotations

As touched on earlier, the same drawing may have parts that were completed at different dates, revisions added by different personnel, and different approval signatures. Drawings in a plant are "living documents"—as the plant changes, expands, and starts up new sections, the drawings must be updated to reflect these changes. Version control is an important factor when working with documents and drawings.

Revision dates are indicated on documents and drawings to show changes made over time. They indicate which copy of the same drawing or document is the most current (as-built) version.

As illustrated in Figure 3-1, revisions may be identified alphabetically (A, B, C, etc.) or numerically (1, 2, 3, etc.). Along with the revision, additional information indicates what changes were made and by whom. For example, changes to a drawing are typically identified by a cloud with the revision letter or number inside a triangle (see Figure 3-3) near or inside the cloud. Revisions to documents are often shown in a section of the document with a date and the reason for the change. For documents (not drawings that have multiple pages), dates on the cover page and header and footer of each page often indicate revision status and who made or approved the change.

If you have questions, you should contact the person whose initials are indicated in the revision section. If you have a question regarding whether certain changes have been approved, you should check the approvals section of the document.

Symbols

Many companies use industry standards for equipment, instrumentation, and graphic elements. Including a set of standardized graphical symbols in a drawing helps maintain consistency, aids in training and understanding by all personnel using the drawing, and—during planning, design, and start-up—helps ensure unambiguous communication between all parties. Usually, the first drawing in a set of P&IDs is the legend sheet that shows which symbols are used in the drawing set and what they mean.

Relevant Standards and Recommended Practices

American Society of Mechanical Engineers

- ASME Y14.100-2017, *Engineering Drawing Practices*
- ASME Y14.24-2012, *Types and Applications of Engineering Drawings*
- ASME Y14.35-2014, *Revision of Engineering Drawings and Associated Documents*

Institute of Electrical and Electronics Engineers

- IEEE 488.1-2003 (R2009), *IEEE Standard for Higher Performance Protocol for the Standard Digital Interface for Programmable Instrumentation*

American National Standards Institute/International Society of Automation

- ANSI/ISA-5.06.01-2007, *Functional Requirements Documentation for Control Software Applications*
- ANSI/ISA-12.02.02-2014, *Recommendations for the Preparation, Content, and Organization of Intrinsic Safety Control Drawings*

International Society of Automation

- ISA-RP60.4-1990, *Documentation for Control Centers*
- ISA-5.2-1976 (R1992), *Binary Logic Diagrams for Process Operations*
- ISA-5.4-1991, *Instrument Loop Diagrams*
- ISA-5.5-1985, *Graphic Symbols for Process Displays*
- ISA-5.3-1983, *Graphic Symbols for Distributed Control/Shared Display Instrumentation, Logic, and Computer Systems*

Drawing References

Often a drawing for a part of the process cannot fit on one page, particularly in the case of process flow diagrams (PFDs) that continue throughout the entire plant. For example, a reference between two PFDs will show the continuation to the other drawing from the originating drawing (and vice versa). These references may be in the form of text at the bottom of a drawing or within a line or stream (identified as A776-O-878-001 in Figure 3-4) directing the reader to another drawing with this number.

References can be made to different types of drawings, too. For example, an electrical drawing may refer to a layout drawing for a panel. Where the "callout" from the electrical drawing to the layout drawing is made, the layout drawing number will be

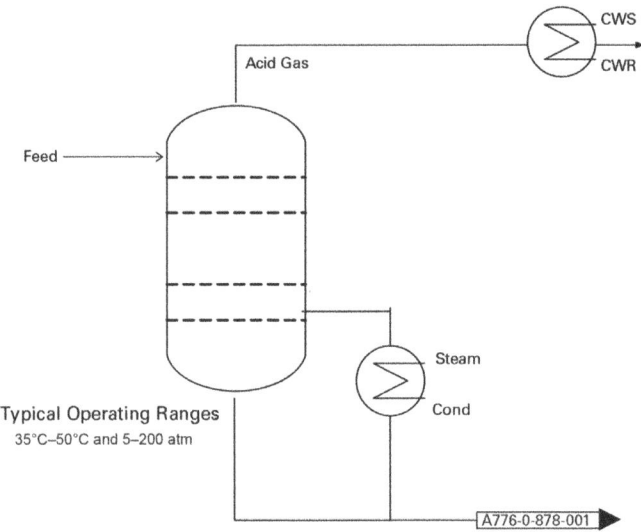

Figure 3-4. PFD referencing another PFD.

indicated. References to documents may also be made. For example, a P&ID may refer to a process description or a standard operating procedure (SOP). It is important to ensure these references are correct and up to date among all drawings as they comprise a complete package (road map) of the process and project.

3.3 Document Locations

Software on the PIN enables people to view start-up drawings and other documents from computers throughout the plant. Typically, the software offers view-only (also known as *read-only*) access, so viewers are unable to make changes. This is a critical feature and part of a good EDMS. The system administrator must determine early in the project who has read-only access and who has read/write access. This is especially true for contract employees because of security concerns.

So far, we have talked mostly about electronically stored drawings. Other documents that may be provided electronically by the PIN are Safety Data Sheets (SDSs) and SOPs. An SDS may be scanned into the PIN if the manufacturer has not provided an electronic version and because it is a document that would not be changed by the local plant. An SOP, however, may be changed often but by a limited number of people. SOPs are often created in a word processing program, but where data and checklists are required, it may be in spreadsheet format.

The Internet and company intranets enable companies to provide start-up documents in a format that multiple individuals using web browsers can quickly view. By putting these documents on a network, personnel can access current documents at different plants owned by the company. This is more convenient than handing out copies, which are difficult to maintain as current.

If the documents are not on the plant's network, their availability may be more limited. Drawings and other documents are often located at their point of use, with their specific locations being up to the individual plant. The following locations are typical:

- Control room
- Electrical shop
- Instrument and electrical (I&E) shop
- Motor control center or input/output (I/O) room
- Engineering, Environmental, and Purchasing departments
- Maintenance shop

As mentioned earlier, documents may be available electronically (via the PIN or PC network), as hard copies (printouts), or if the process is an old one, as original hand drawings. Originals should not be removed from their locations except to be copied for use in the field.

3.4 Commissioning and Start-Up Drawings and Documents

You will commonly encounter several different types of documents and drawings during commission and start-up. This section provides a brief summary of the most important types.

3.4.1 Gantt Charts

One of the first documents developed for a project, and therefore crucial to plant start-up, is the project Gantt chart (see Figure 3-5), which is often referred to as a *critical path* diagram. This document provides a sequence of activities that must be completed by certain dates for the entire project to be completed on schedule, and is most often referred to as the *project schedule*.

A Gantt chart is a type of bar chart. While the project is being designed and throughout the start-up process, the Gantt chart is a graphical means used to depict the activities during the total project, the time each activity will take to complete, and

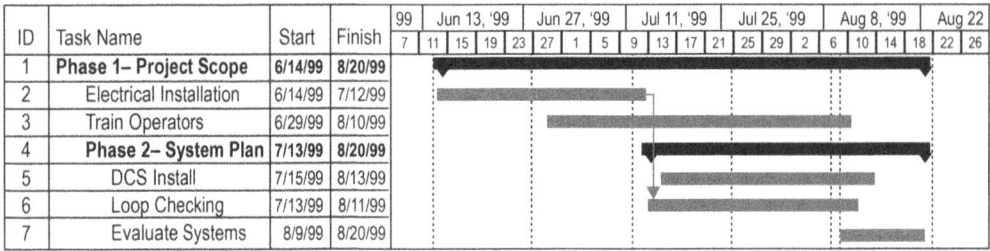

Figure 3-5. Gantt (or *critical path*) chart.

how the activities interact with each other (relationships). There are several software applications that enable users to create effective Gantt charts. Two popular ones are Microsoft Project and Primavera P6.

The Gantt chart for the instrumentation and automation part of the project a CST would be involved with may be broken up, for example, into groundwork, equipment placement, process automation system (PAS) installation, electrical and instrument installation, calibration, loop checking, training, and start-up. These activities are shown on the chart as horizontal bars that represent the amount of time the activities will take to complete. The chart also shows the relationship between activities, especially if one activity must be completed before another can start.

Figure 3-5 shows that the task Electrical Installation will commence the second week of June and last for 1 month. This task must be completed before loop checking can begin during the second week of July. Other activities, such as operator training and PAS installation (distributed control system—DCS Install), will also occur during this time frame.

The person responsible for determining how long the various tasks should take and for keeping track of their progress is the PM. If the schedule slips or there is a delay in part of the project, the Gantt chart is updated (by the PM and/or planner/scheduler start-up team members) to reflect the change and the interaction of the affected activities.

3.4.2 Functional Specifications

Also commonly called *functional requirements*, a *functional specification* is a document that details a product's (equipment or software) or system's intended capabilities, appearance, and interactions with users. This specification should provide the product's or system's capabilities to meet user requirements or objectives. It is a blueprint for the design of the product or system. The CST, for example, may want to review a functional specification for a compressor control system to understand how the control system associated with it is intended to automate this part of the plant process.

3.4.3 PFDs

A PFD is a diagram used to indicate the general flow of plant processes and the functional relationships between the process and major pieces of equipment. Not much detail is shown. It is not to scale, nor does it depict a plant layout as a *general arrangement* (GA) diagram does (see Figure 3-6). The PFD shows only major pieces of equipment, and piping runs are represented in shorthand fashion—as straight lines and not drawn to scale—rather than as detailed runs between equipment depicted on a P&ID.

The PDF gives a *functional overview* of the process. One difference between a PFD and a P&ID is that a PFD often includes process information, such as material and product streams, flow rates, temperatures, and pressures. In addition, a PFD is usually set up to provide a complete mass and energy balance for the process. A simplified example of a PFD is shown in Figure 3-6.

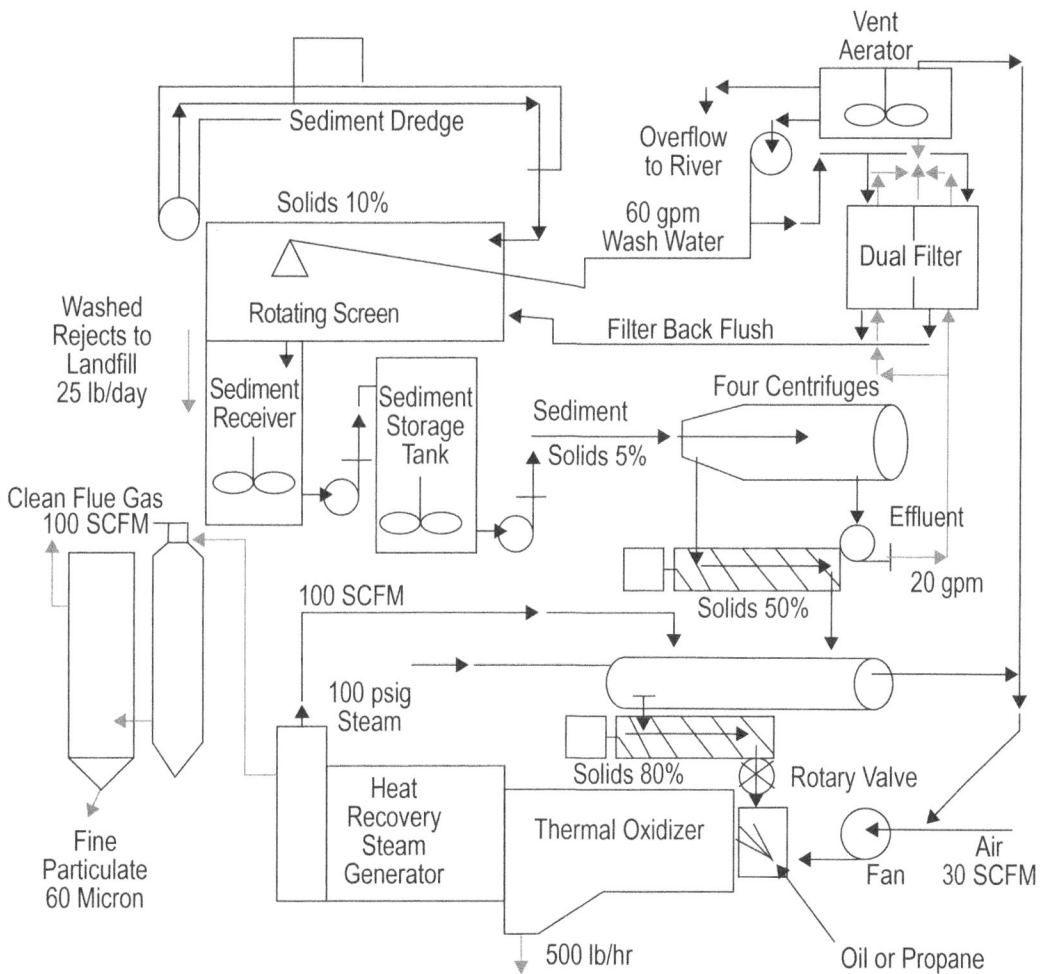

Figure 3-6. General PFD.

The information contained in a PFD can be used by the CST for replacing instrumentation such as flowmeters, for calibrating, and for understanding any safety implications of the process. It is also an important document for understanding what the plant is going to produce and how it will produce it, because it presents an overall picture of the process in a few pages.

A PFD is typically drawn for a single process unit (e.g., distillation column) and includes the following:

- Process piping
- Major pieces of equipment
- Control valves and other major valves
- Connections with other systems
- Major bypass and recirculation streams
- Operational data (temperature, pressure, mass flow rate, density, etc.), often by stream references to a mass and energy balance
- Process stream names
- Some instrumentation; for example, a level control loop to indicate which flow is manipulated to maintain a vessel level

A PFD generally does *not* include:

- Pipe classes or piping line numbers
- Detailed process control instrumentation (sensors and final elements)
- Minor bypass lines
- Isolation and shutoff valves
- Maintenance vents and drains
- Relief and safety valves
- Flanges

PFDs are usually used by engineers and environmental personnel. An example of their use by environmental personnel is calculating the mass flow of plant effluents (discharges to the air, ground, and water) to ensure a plant's environmental compliance. Most of the materials that enter the process (inputs) go out as product, but some may become waste. Engineers use these drawings to calculate production rates, size equipment, scale up plants to increase production, or design a plant from a pilot plant.

CSTs may also find it beneficial to peruse such drawings. For example, you may use the information shown in a PFD to understand the overall process.

Mass (or Material) and Energy Balance

Mass balance information is usually included on the PFD. This information provides a means of tracking materials in and out of the process mathematically. It is set up so the sum of the mass flows of chemicals, compounds, or other materials *into* the process equals the sum of their mass flows *out* of the process. These calculations may be complicated by any reactions that occur in the process. This is because any chemical conversion or material change of state (e.g., evaporation) affects the quantities in streams shown on these drawings. The chemical compositions of all ingredients and intermediate and final products must be known to perform an accurate mass balance.

Figure 3-7 and Equations 3-1 through 3-3 comprise an example of a mass balance calculation. Such a calculation may be done for one component of a process (e.g., ammonia) or for all mass flows associated with the product being produced. All these flows must be taken into consideration when calculating production quality and yield, determining material handling requirements, and dealing with pollution concerns. In turn, these flow rates, pressures, and temperatures are important when specifying and installing instrumentation during a new plant project, as well as when troubleshooting and replacing equipment in an existing plant.

Figure 3-7 indicates that 1000 lb of a substance containing two components, X_A and X_B, enter the process in a ratio of 20% X_A and 80% X_B. A quantity of 700 lb exits with a composition of X_A and $0.3X_B$ and $W(0.1X_A$ and $?X_B)$.

The unknowns, X_A in the 700 lb stream and X_B in stream W, can be determined mathematically using two simultaneous equations:

$$1000(0.2X_A + 0.8X_B) = 700(?X_A + 0.3X_B) + W(0.1X_A + ?X_B), \text{ where } W = 300 \text{ (ideally)} \quad (3\text{-}1)$$

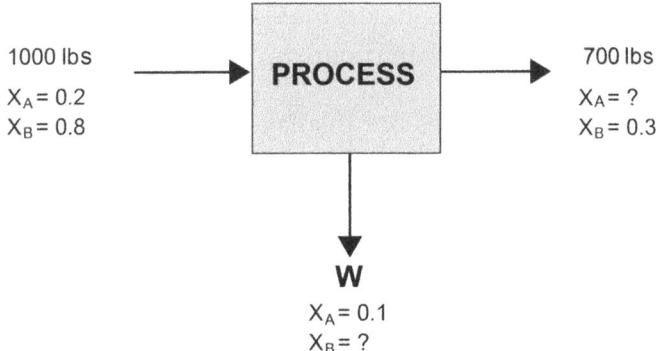

Figure 3-7. Mass (material) balance.

and

$$X_A + X_B = 1.0 \tag{3-2}$$

Energy Balance

Similar to the mass balance, the sum of all energies (Σ E) going into a process must equal the sum of the energies going out.

An energy balance may be complicated by the different energy values that must be considered for varying temperatures and pressures, as well as by any chemical reactions that may occur. An energy balance is a means of calculating the energy requirements of a process for heating, cooling, and reaction.

$$\Sigma E_{IN} = \Sigma E_{OUT} + \Delta \Sigma_{REACTION} \tag{3-3}$$

3.4.4 P&IDs

The P&ID (see Figure 3-3) is the primary schematic depicting a process, typically the start-up process. Of the drawings mentioned in this chapter, the P&ID is the most widely used and forms the basis for many of the other drawings discussed. P&IDs play a significant role in the maintenance and modification of the processes they describe, because it is critical to record and display the physical sequence of instrumentation, equipment, and systems, as well as how these systems connect.

During the design stage, the P&ID provides the basis for the development of system control schemes and enables safety and operational investigations such as hazard and operability studies and process reviews.

A P&ID shows major equipment, piping, and instruments, and all pneumatic, electrical, hydraulic, magnetic, and logical (software) connections between them. It does not attempt to accurately depict piping runs, but it does show details pertaining to instrumentation, valve type, piping specifications, special equipment, and tie-in points. This drawing often includes logical interlock information as well as notes and bubbles denoting changes and work in progress.

P&IDs also show whether pipes are insulated, and often indicate where a pipe goes through a major wall, floor, or roof. Other data, such as materials of construction, pump size, or head and impeller size, may be shown in blocks at the bottom of the P&ID. The legend, which is described and illustrated later in this section, is the guide that tells you how to read the information on a P&ID.

As a CST, the information contained in the P&ID is important to you when you are involved in rebuilding valves or replacing flow rate or pressure measurement elements in process lines, but the P&ID as a whole is important to you because it depicts the relationship between the PAS and field equipment (process control). The P&ID, loop sheets, and/or spec sheets are important leading up to and during start-up.

Most companies use ISA standard symbols on the P&ID. This is important because standard symbols enable those who relocate to another plant to aid in start-ups there and to understand a new system—their knowledge is transferable. In addition, design and engineering firms, which often produce P&IDs for industry, can produce a consistent product when they abide by ISA standards.

Usually, the first drawing in a P&ID set depicts the standard nomenclature and symbols used for the set of drawings. This first page is typically called a *legend sheet* (see Figure 3-2). Figure 3-2 also shows the different types of valves, indicated by their P&ID symbols.

In addition to those just mentioned, items typically shown on a P&ID include the following:

- All instrumentation, including local pressure gauges, sensors, and transmitters, as well as valve controller and instrument designations
- Mechanical equipment with names and numbers
- All valves and their identifications
- Process piping, sizes, and identification
- Miscellaneous—vents, drains, special fittings, sampling lines, reducers, and increasers
- Permanent start-up and flush lines (as opposed to temporary hoses)
- Flow directions
- Interconnection references
- Control inputs and outputs, and interlocks
- Automation and control system inputs
- Identification of components and subsystems delivered by others

3.4.5 General Arrangement Drawings and Plot (Location) Plans

General arrangement (GA) drawings help you physically locate the process equipment you will be working on. Often called *plot plans*, these drawings (see Figure 3-8) comprise an accurate architectural rendering of the plant. They are drawn to scale, and each GA shows different views so you can look down from above to see plant equipment and elevation drawings, including the building structures and major pieces of equipment.

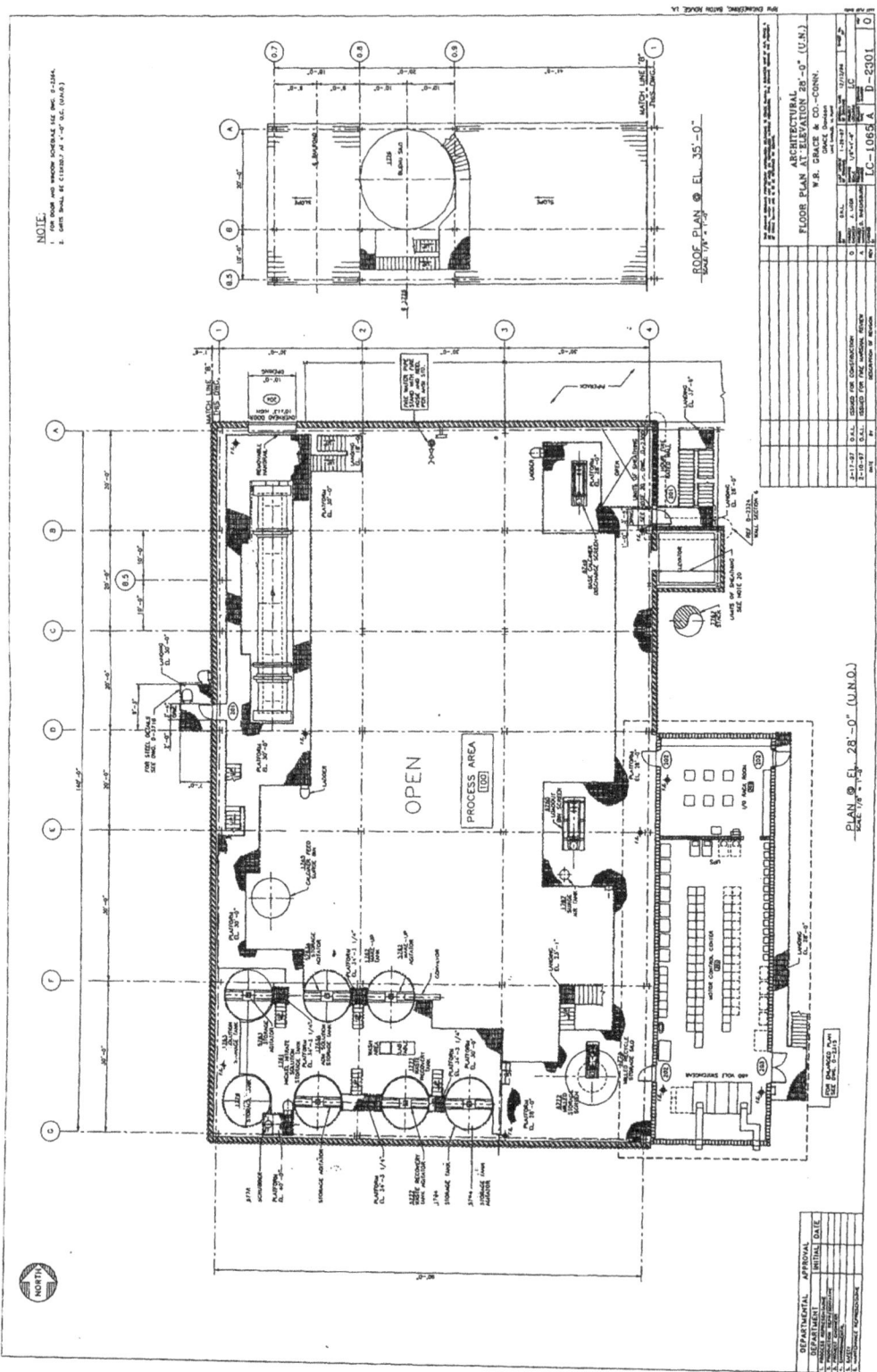

Figure 3-8. GA drawing.

Source: *Courtesy of Grace Davison, Lake Charles, LA.*

A difference in elevation may be important, for example, when you are calibrating certain instrumentation. Some plants use locations to name their equipment as well. For example, a prefix or suffix may be added to indicate location, such as N for *North*.

Often new instrumentation associated with the start-up is shown on a GA drawing along with elevation. GA drawings will help you find the new instrumentation in the process areas when you must install, calibrate, replace, or troubleshoot them during the start-up.

3.4.6 Instrument Specification Sheets

Companies normally have sets of instrument specification "spec sheets" for each type of instrument. That means there is one set for each model of temperature transmitters, pressure transmitters, flow transmitters, and valves. Each spec sheet contains a list of requirements that must be met when manufacturing and assembling the instrument and for testing its attributes or functions, especially during calibration and loop check.

Spec sheets present basic instrumentation information and may not include all necessary engineering data or definitions of application requirements. ISA standard spec sheets (in ISA-20) are intended to cover the most commonly used instruments and control valves. New forms are added with each general revision of the standard. While the types of instruments described by the ISA standard are more common to the process industries, the forms also prove useful in other areas if special requirements are defined. Some forms consist of a primary sheet and a secondary (tabulation) sheet. The primary sheet may be used by itself to specify a single instrument or to specify general requirements for a series of similar instruments that are then listed on the secondary sheet.

The purpose of the ISA standard spec sheets is to promote uniformity in instrument specifications, in both content and form. Because of the complexity of present-day instruments and controls, it is desirable to have some type of specification form to list pertinent details for use by all interested parties. General use of these forms by users and manufacturers offers many advantages, such as the following:

- Assists in preparing a complete specification by listing and providing space for all principal descriptive options
- Promotes uniform terminology
- Facilitates quoting, purchasing, receiving, accounting, and ordering procedures by uniform display of information
- Provides a useful permanent record and means for checking the installation
- Improves efficiency from the initial concept to the final installation

Many spec sheets have headings to permit the user to add the company name, plant location, trademark, or specific project data (see Figure 3-9). With an ISA standard spec sheet, an instruction sheet is usually provided for each form to explain the terms used and the intended procedure. The instructions are keyed to the form by reference to the line numbers. The ISA20 committee has minimized dependence on the instruction sheet because the forms are frequently reprinted and used without the instructions. The explanation is omitted where the meaning is felt to be obvious. Instrument specifications may be prepared using computerized systems. The format of such specifications may be modified to be compatible with computer software, such as SmartPlant (INtools) by Hexagon PPM, formerly Intergraph.

DIFFERENTIAL PRESSURE INSTRUMENTS

NO	BY	DATE	REVISION	SHEET	OF
				SPEC. NO.	REV.
				CONTRACT	DATE
				REQ. - P.O.	
				BY / CHK'D	APPR.

	1	Tag Number	Service
GENERAL	2	Function	Record ☐ Indicate ☐ Control ☐ Blind ☐ Transmit ☐ Integ ☐ Other
	3	Case	Mfr Std ☐ Nom Size Color: Mfr Std ☐ Other
	4	Mounting	Flush ☐ Surface ☐ Yoke ☐ Other
	5	Enclosure Class	General Purpose ☐ Weather Proof ☐ Explosion Proof ☐ Class
			For Use in Intrinsically Safe System ☐ Other
	6	Power Supply	117V 60Hz ☐ Other ac dc ☐ Volts
	7	Chart	12 in. Circ. ☐ Other Range No.
	8	Chart Drive	24 hr. ☐ Other Elec. ☐ Spring ☐ Other
	9	Scale	Type Range: 1 2 3
XMTR	10	Transmitter Output	4-20 mA ☐ 10-50 mA ☐ 21-103 kPa (3-15 psig) ☐ Other
			For Receiver, See Spec Sheet
CONTROLLER	11	Control Modes	P = Prop (Gain), I = Integral (Auto Reset), D = Derivative (Rate), Sub: s = Slow, f = Fast
			P ☐ PI ☐ PD ☐ PID ☐ I_f ☐ D_f ☐ I_s ☐ D_s ☐
			Other
	12	Action	On Meas. Increase Output: Increases ☐ Decreases ☐
	13	Auto-Man Switch	None ☐ Mfr Std ☐ Other
	14	Set Point Adj.	Manual ☐ External ☐ Remote ☐ Other
	15	Manual Reg.	None ☐ Mfr Std ☐ Other
	16	Output	4-20 mA ☐ 10-50 mA ☐ 21-103 kPa (3-15 psig) ☐ Other
UNIT	17	Service	Flow ☐ Level ☐ Diff. Pressure ☐ Other
	18	Element Type	Diaphragm ☐ Bellows ☐ Mercury ☐ Other
	19	Material	Body Element
	20	Rating	Overrange Body Rating psig
	21	Diff. Range	Fixed ☐ Adj. Range Set At
	22		Elevation Suppression
	23	Process Data	Fluid Max Temp. Max Press.
	24	Process Conn.	1/2 in. NPT ☐ Other
	25	Alarm Switches	Quantity Form Rating
	26	Function	Meas. Var. ☐ Deviation ☐ Contacts To on Inc. Meas.
	27	Options	Pressure Element ☐ Range Material
			Temp. Element ☐ Range Type
			Filt-Reg. ☐ Sup. Gage ☐ Output Gage ☐ Charts
			Valve Manifold
			Cond. Pots ☐ Adj. Damp ☐ Integral Sq. Rt. Ext. ☐
			Integrator
			Other
	28	Mfr & Model No.	

NOTES:

 ISA FORM S20.20a

Figure 3-9. Spec sheet.

Relevant Standard

International Society of Automation

- ISA-20-1981, *Specification Forms for Process Measurement and Control Instruments, Primary Elements, and Control Valves*

3.4.7 Loop Diagrams (Loop Sheets)

Loop diagrams, also known as *loop sheets* (see Figure 3-10), are among the drawings that a CST uses most often. Each loop diagram schematically represents a complete hydraulic, electric, magnetic, pneumatic, or logical circuit. Loop diagrams often contain information pertaining to the instrument signal type and measurement range for analog signals, instrument manufacturer, field junction box designation, marshaling panel connection (where applicable), PAS, PAS termination, tag name on the PAS, control system (DCS/basic process control system—BPCS, safety instrumented system—SIS, or the third-party packaged system—TPPS interface), and a description of the operator interface.

For digital signals, these drawings show the state of closure of each electrical contact and how the contacts are wired and fused.

Most companies have standards for their loop sheets. For example, a loop sheet often depicts a single loop (e.g., field devices—transmitter, valve, and limit switches)

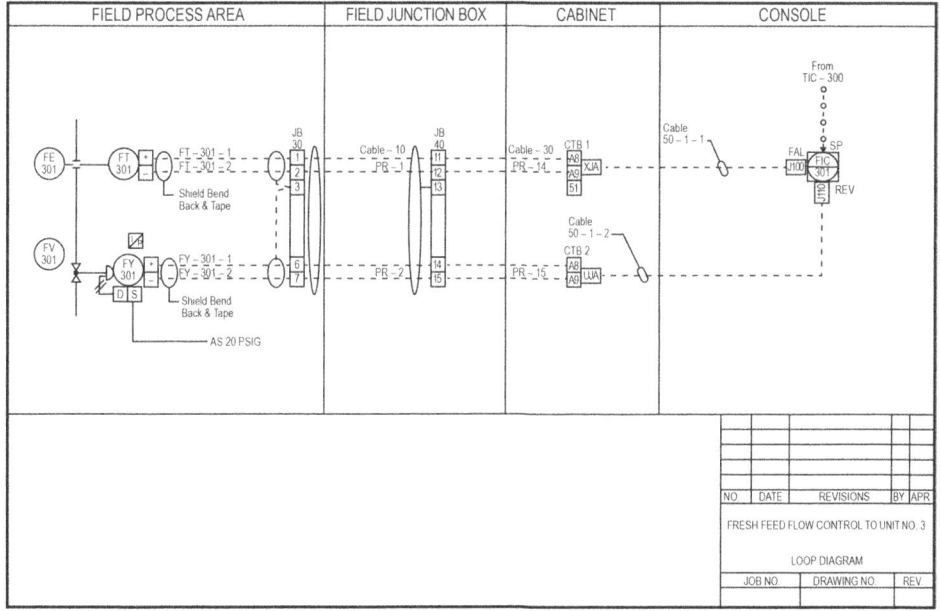

Figure 3-10. Loop diagram/loop sheet.

on one page. To read and use loop sheets, one must understand DCS and programmable logic controller (PLC) terminations, as well as electrical assembly. As with P&IDs, these drawings should have a standard format that is used throughout the plant.

3.4.8 Loop Check Log Sheets

The loop check log sheet (see Figure 3-11) might only be used during commissioning, and not at all plants. (Figure 3-11 is a sample list of loops to be checked.) Its function is to list all I/O signals that must be checked out prior to start-up. Usually, a master copy is maintained to avert any confusion about what has been checked out and what has not. This document contains each of the tag names on the loop diagrams, a description of functional information, problems encountered, work that must be completed, and a column for signing off on completion of the loop check.

To perform loop checks, you might be stationed in the plant or in the I/O room while the operator in the control room "strokes" (moves) a valve or starts and stops one or more pieces of equipment. The operator then looks for an indication on his or her panel or PAS display (human-machine interface—HMI) that matches the actual state of the valve or piece of equipment. If the loop does not work properly, you must fix it and then try the loop again with the operator's help. You may have to mark up

Loop Check Log Sheet							
					Project LC-1085 Control Room		
Valves and Motors: Toggle/stroke output to each state (e.g., OPEN/CLOSED; START/STOP) and verify matching feedback. *Transmitters and Control Valves: Simulate field signal (4–20 mA 5-point check). Stroke valves and check valve position.							
Tagname	Output (OP)	Field Signal* (PV)	Feedback (PV)	Notes (include problems encountered/solutions)	Initials	Date	OK
1. UV64555							
2. FC64554							
3. PC64118							
4. PC64191							
5. LI12500							
6. LI13540							
7. FI15400							
8. FI1456705							
9. SI50402							
10. TC34659							
11. TI34321							
12. PDI86543							
13. VI23450							
14. LC63999							
15. DI53555							
REF: P&ID D-1715							

Figure 3-11. Example of a loop check log sheet to be used during loop check activities for a particular set of loops.

(redline) the loop sheet if something was drawn in error. It is important for you to communicate any changes through the plant MOC process so that revised drawings can be generated that properly show how the loop works. See Chapter 5 for a further discussion of the MOC process.

As an example of such a change, perhaps the range of an instrument must be adjusted to match the actual conditions of the loop. For more information, see *Loop Checking: A Technician's Guide* by Harley M. Jeffery.

Sometimes loop check log sheets are not used, and the CST is given a stack of loop folders containing all the necessary documents to perform loop and function checks. When the CST is done with the folders, he or she may receive another stack to work on. Regardless of whether a loop check log sheet or stack of loop folders is used to complete the loop check process, someone must track that all project instruments and signals have been checked out.

3.4.9 Calibration Data Sheets

A calibration data sheet (see Figure 3-12) must be created for each process instrument and must document the identity of the person who performed a given calibration, that it was completed, and that the specification tolerances were achieved. The sheet also must be dated. Calibration data sheets are primarily used to do the following:

- Determine the locations at which scale graduations are to be placed
- Adjust the output and to bring it to the desired value, within a specified tolerance
- Ascertain the error (comparison of the instrument output reading against a standard)

The CST may calibrate an instrument in the maintenance shop, on a bench, or in the field. Some instruments can be purchased precalibrated, but many companies still have a CST check each instrument in the shop and sometimes again in the field. Full loop calibration requires the instrument to be in the loop in which it will be operating when the unit/plant/system is operational. This is standard practice in many plants and is often required by regulation. For more information, see *Calibration: A Technician's Guide* by Mike Cable.

3.4.10 Installation Details

An installation detail (see Figure 3-13) shows how an instrument is to be mounted in the field. It is often drawn to scale and generic for the type of equipment being installed, so it applies to different equipment if it is referenced as such. One example of

INSTRUMENT CALIBRATION SPECIFICATION LOG							
INSTRUMENT	TECHNICIAN	M.R	ORIG. DATE	OCT - 95	REV. #	BY	DATE

TAG:	DT12034		
SERIAL #	9.53202E + 12		
MANUFACTURE:	HONEYWELL		
MODEL:	CMX201-DFBA377		
PLANT:	FCC-NEW		
SERVICE:			
PARAMETERS:	VALUE		
ID	DT12034	PHASE 1 TC = 0	
UNITS PV1 =	GPM	PHASE 2 = 0.3	
PV 2 =	S.G. @60° F	PHASE 2 TC = 0	
PV 3 =	° F	PULSE = N / A	
PV 4 =	% SOLIDS	LIMIT SWITCHES = N / A	
L.R.V PV 1 =	0 GPM	DETECTOR PIPE SIZE = 50 MM 2"	
PV 2 =	0.95 S.G. @60° F	PIPE ZERO 1 = 127	PIPE ZERO 2 = 127
PV 3 =	0° F	K FACTOR 1 - .9094	K FACTOR 2 = .9094
PV 4 =	% SOLIDS	FLOW DIRECTION - FORWARD FLOW	
U.R.V PV 1 =	150 GPM	OUTPUT POLARITY - ACTIVE FORWARD	
PV 2 =	1.3 S.G. @60° F	BPA = UNIPOLAR	
PV 3 =	120° F	FRONT END FILTER = MEDIUM	
PV 4 =	100% SOLIDS	ANALOG OUTPUT 2 = DENSITY	
CONFIGURATION		LOW FLOW CUTOFF	
FLOW PV 1 =	VOLUME FLOW	LOW FLOW MODE = FLOW UNITS	
BBL CONVERSION	OIL = 42 GAL	LOW FLOW LIMIT = 4.5 GPM	
% SOLIDS DENSITY	PHASE 1 = 1.00	LOW FLOW HYST = 3.0	

AUXILIARY INPUT = NOT ASSIGNED   DDL = 0.3
POSITIVE ZERO RETURN = INACTIVE
        RANGE   DUAL RANGE = MEAS.
MEASURING RNG 2 = 100 GPM
OPTIONS = % SOLIDS - INSTALLED
LICENSE CODE = A/O 2 - NOT INSTALLED
2 LIMIT SWITCHES NOT INSTALLED
TOTALIZER = READ RESET FUNCTIONS
FAIL SAFE = HIGH
AUTOSETUP = NOT IN PROGRESS N / A
VERSION INFO = SENSOR SWVER 02.01.01
STORES CODE #

Figure 3-12. Calibration data sheet.

an installation detail is the required distance, measured in pipe diameters, of a flowmeter from a turn (or other source of turbulence) in the pipe for the meter to measure accurately. Another example is the position (horizontal or vertical) in which a flowmeter should be installed in a line. You should have these drawings and instructions when you mount new instruments for a project or replace a defective instrument.

3.4.11 Other Documents and Drawings

Manufacturers' Information Documents

The vendor typically provides the plant with several copies of manufacturers' information documents. The vendor considers these documents to be deliverables for the company that has purchased the piece of equipment or system. There are multiple copies of the

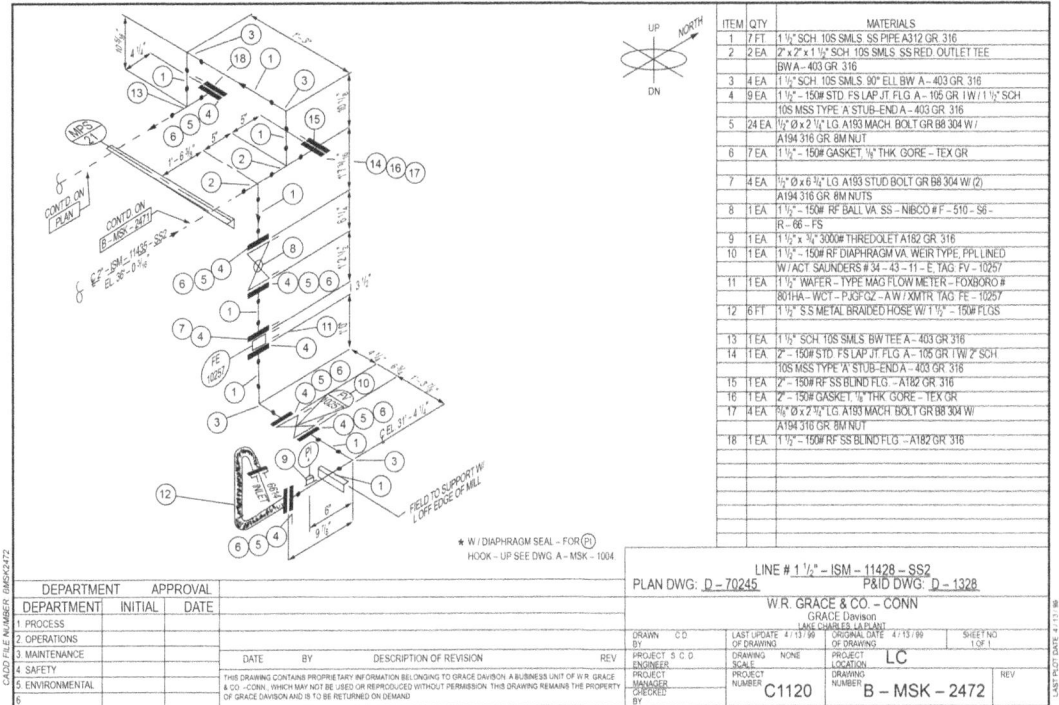

Figure 3-13. Installation detail.
Source: Courtesy of Grace Davison, Lake Charles, LA.

documents, and one is typically available in the maintenance shop. This documentation is important to you because it contains information pertaining to equipment tolerances and operation so, for example, you can calibrate field instruments to measure properly or understand the PLC TPPS program (when applicable) needed to make the equipment run. Manufacturers' documents may also include lubrication schedules for rotating equipment and explanations of how to use and maintain the equipment. Examples of the types of equipment that come with manufacturers' information documents are packaged units such as air dryers and deionized water units, compressors, pumps, and valves.

These documents differ from specification sheets. A *spec sheet* is usually a one-page document that summarizes information or data for a single instrument. Spec sheet information is important for design (size, capacity, temperature limits, material of construction, etc.) and calibration (range). *Manufacturers' information documents* are often large documents, even books or binders containing detailed information (including instructions and drawings) pertaining to the equipment purchased.

Electrical Wiring Diagrams

Electrical wiring diagrams (see Figure 3-14), sometimes called *one-lines*, may come from the engineering design firm that did the plant design, but they also come with

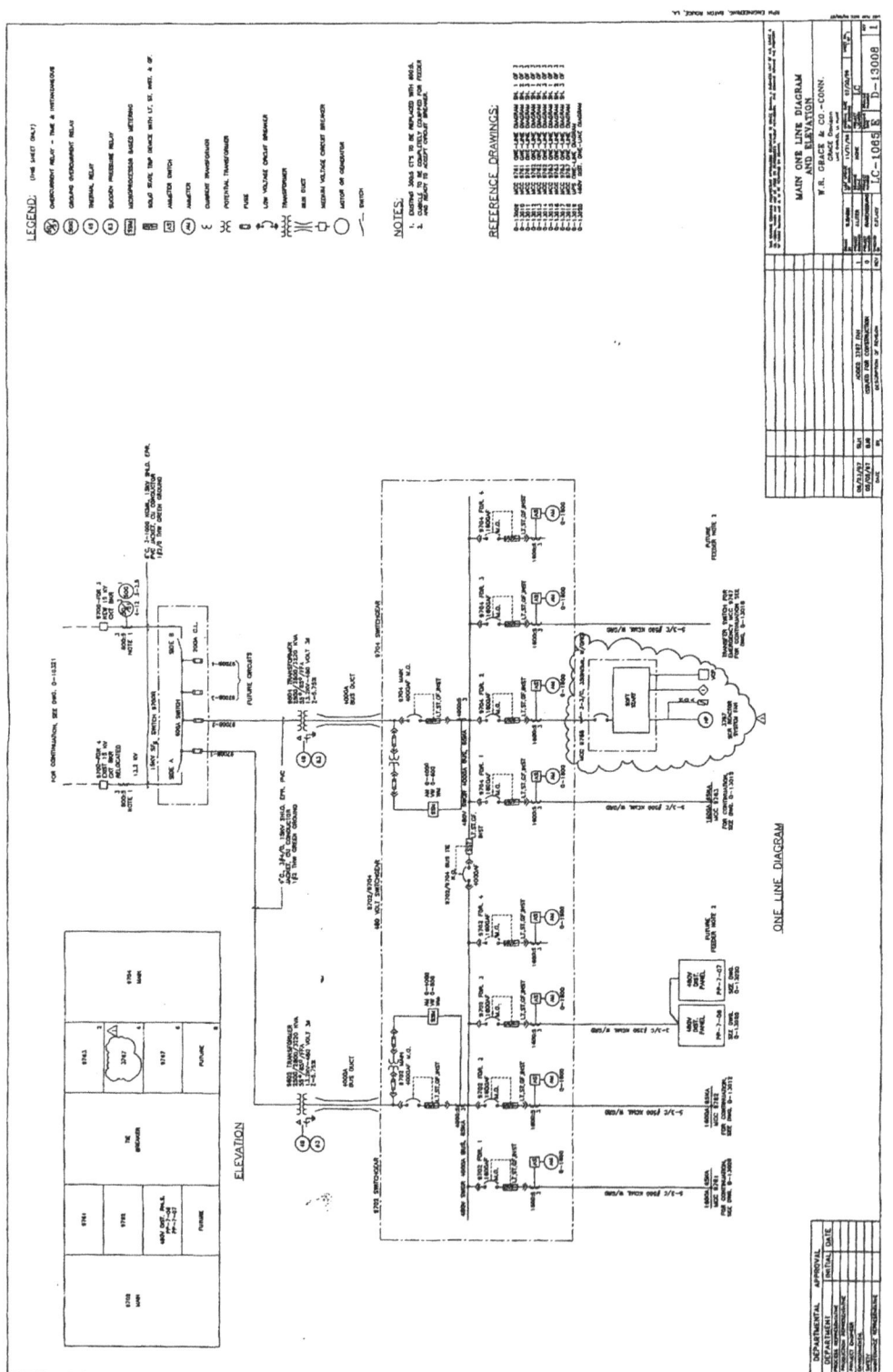

Figure 3-14. Electrical wiring diagram.
Source: Courtesy of Grace Davison, Lake Charles, LA.

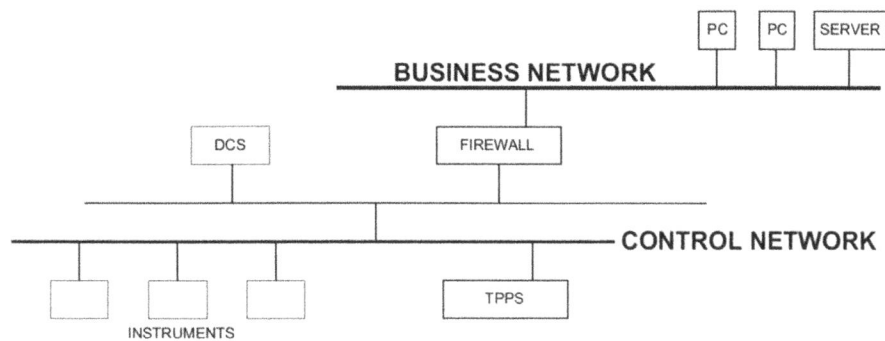
Figure 3-15. Network topology diagram.

TPPSs such as burner management systems, weigh feeder systems, and pneumatic conveying packages. If these systems cause trouble during start-up and troubleshooting, your understanding of electrical drawings will greatly benefit the start-up effort. Difficulties arise from the fact that these packaged systems must be integrated with existing plant wiring and systems.

Network Topology Diagrams

Following is a generalized block drawing indicating the relationship between the industrial control and business networks. This topology drawing indicates the way in which these systems' parts are interrelated or arranged. Other drawings used during the project that fit into this drawing category show more detailed system parts and relationships, often limited to each network (control system or business network), including computers, peripheral devices, assorted modules, converters (e.g., coax to fiber), and cables. To assist effectively during check-out and subsequent start-up, these diagrams must include good documentation indicating component numbers, labeling, and wire/cable numbers to enable verification against actual equipment prior to start-up. These drawings, similar to others, must be redlined (and subsequently updated) to indicate as-built status so the actual installation is documented correctly.

Figure 3-15 shows two networks that interface through a security device known as a *firewall*.

Logic Diagrams and Other Control Logic Documentation

Electrical equipment, and therefore electrical wiring diagrams, may interface with PLCs. Consequently, it is helpful for you to know Ladder Logic (see Figure 3-16) and PLC programming. How much you must know should be determined by the project requirements and the role of the CST during start-up. Depending on these requirements and the PLCs used in the plant, you may need additional training, which is usually available from the PLC vendor. Also, depending on the level to which you

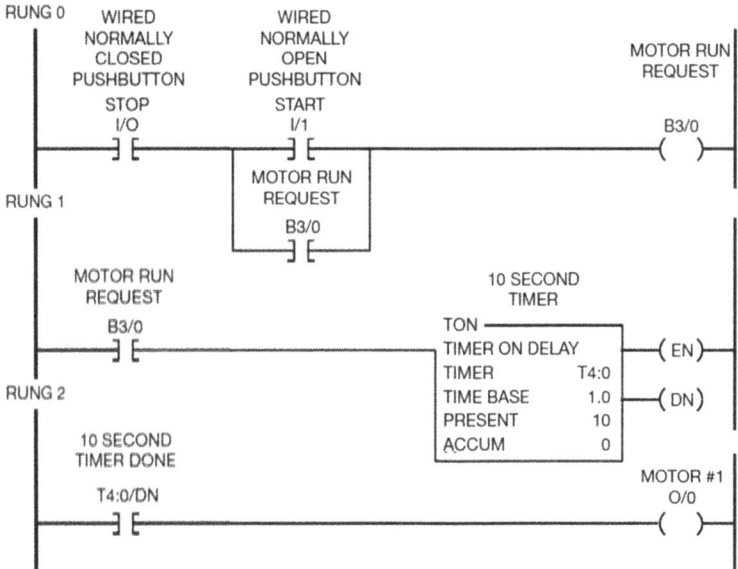

Figure 3-16. Sample of PLC Ladder Logic.

will be troubleshooting connections or writing code for a PLC, you may need to go to the vendor for several courses. A full discussion of Ladder Logic is beyond the scope of this book, but you can learn more about the topic from the many courses vendors offer.

The Ladder Logic depicted in Figure 3-16 is used to turn on a motor after a programmed delay time preset of 10 s.

Other types of logic documentation that you may be exposed to during the start-up are as follows:

- **Process control narratives** – These describe how a process is to be operated in all modes of process control, such as start-up, shutdown, and emergency.

- **Cause-and-effect (C&E) matrices** – These matrices are used to list all the different causes that can be attributed to a specific problem and their effect on the process (see Figure 3-17). The effect is often to perform interlock (shutdown logic) checks.

- **Control system logic** – An example is the process control and automation software often programmed by a control systems engineer or system integrator to perform process control.

Each company may have a standard for how control narratives, C&E matrices, and control system logic are written. A full discussion of each of these topics is beyond the scope of this book; there are courses you may take to learn more about them.

01-05

Causes:
Tank 101 Level is High High (LSHH-101 = HIHI) OF
Flow 302 is Low Low (FSLL-302 = LOLO)

Effects:
Close Tank 101 Inlet Valve (UV-101) AND
Stop Pump 302 (P-302 = OFF)

Also may be in tabular form:

Interlock	Causes	Effect
01-05	LSHH-101 = HIHI	CLOSE UV-101
	OR	AND
	FSLL-302 = LOLO	STOP P-302
01-06	etc.	etc.

Figure 3-17. C&E matrix examples.

Flowcharts

There are different types of flowcharts because they are used for different purposes. Flowcharts provide a graphical representation of a process or procedure and may include block diagrams, routine sequence diagrams, and general flow symbols (see Figure 3-18). For example, a flowchart can represent a problem or a way to troubleshoot a problem, the flow of data, procedures, equipment, methods, documents, machine instructions, or a sequence of operations. It uses symbols to represent these operations, documents, equipment, and so on. Flow lines are the connecting lines or arrows between the symbols on the flowchart. The flow line leaving a symbol may be singular or multiple, depending on the result or the question asked at the operation (e.g., YES/NO).

SOPs

A plant's Operations department often employs SOPs to train people and to use as checklists for running the plant and handling emergencies. Such procedures provide a step-by-step sequence that describes how to perform various tasks, including the sequence of operations for starting up and stopping equipment safely and efficiently in the plant. SOPs are beneficial because, as the name implies, they ensure a procedure or task is performed in a *standard* (consistent) way. Over time, SOPs may be replaced with automation sequences that the PAS runs so that an operator does not have to perform the steps manually. It is important for you to understand plant SOPs to ensure that your work does not compromise or interfere with the processes they describe. Understanding SOPs also helps you understand how the plant and operators work to make the product. You must adhere to SOPs and to any emergency procedures.

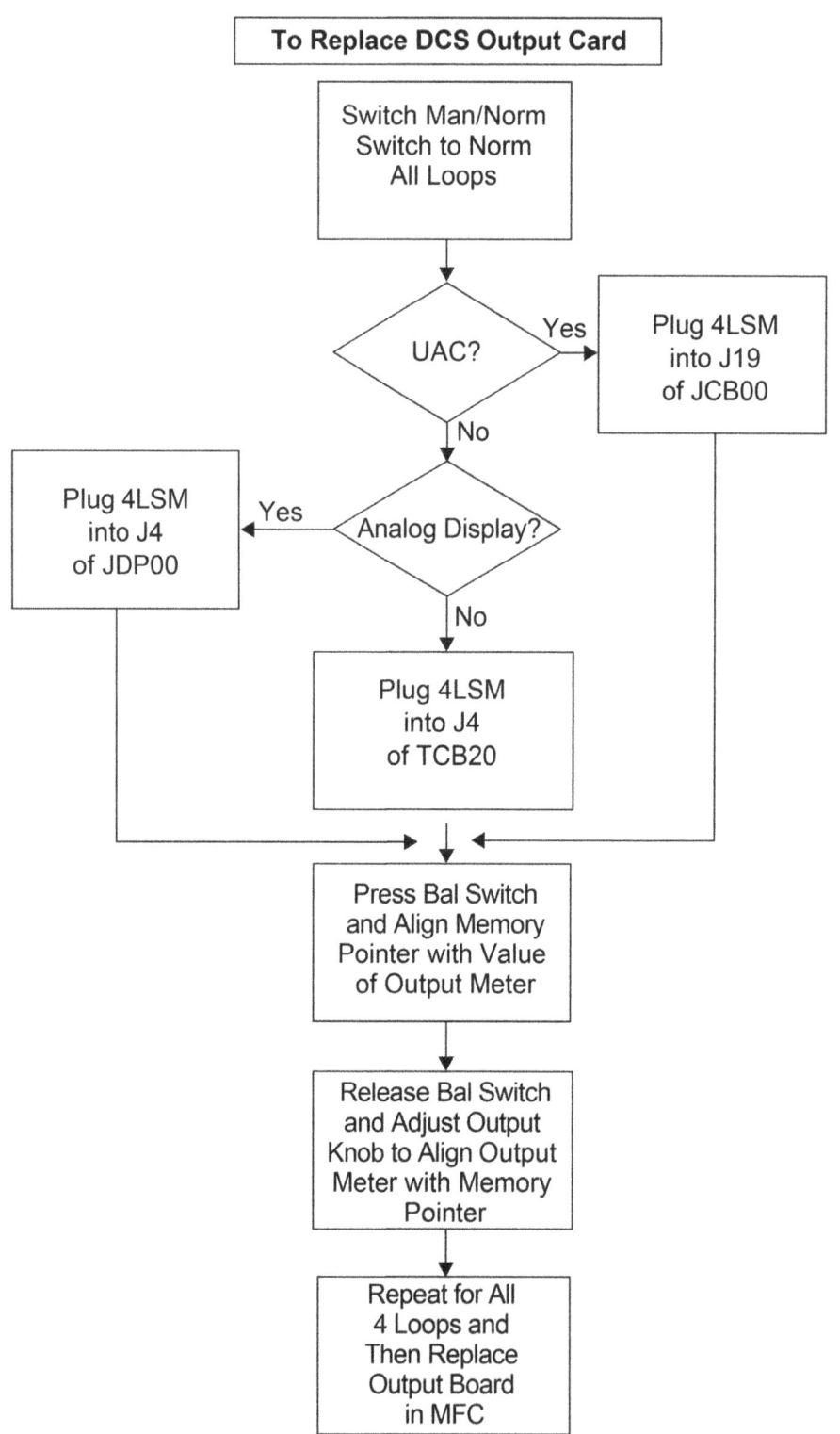

Figure 3-18. Flowchart example.

If you must interrupt a process covered by an SOP to work on equipment related to the process, then you must cooperate with the operator. Doing so ensures that the work stoppage will not negatively affect the plant or plant safety and will occur at a good time for the operator.

Appendix C to US Occupational and Safety Health Administration (OSHA) 29 CFR 1910.119 serves as a nonmandatory guideline to assist employers and employees in complying with the requirements of this section and provides other helpful recommendations and information.

Quality/Inspection Checklist

Sometimes the CST functions as a quality inspector, and other times a third-party quality/inspection contracting company performs the quality/inspection checklist (also known as QA/QC forms) inspections. Typically, the loop must pass the quality inspector's approval and become part of a loop folder before a loop can be loop checked. Any items that do not pass the quality inspectors' examination are added to a punch list indicating something must be fixed and/or addressed.

See the discussion of quality/inspection in Chapters 1 and 5, and the quality/inspection checklist shown in Figure 3-19.

Start-Up Plan

Finally, you may see a pre-startup safety review (PSSR) checklist (see Chapter 8) and an accompanying start-up plan. The PSSR is the final check that will be completed before actually starting up the plant. As opposed to a checklist, the start-up plan is a more detailed description of the activities to be completed. See Chapter 8 for more details on this subject.

Test Documents

Test documents include:

- FAT plan
- FAT test results
- FAT certificate
- Site integration test (SIT) plan
- SIT test results
- Site acceptance test (SAT) plan

```
Loop #: A    1016
Reactor Area

J-Box:    JB202              PID:    901008      Status:
Inspector:                    Area:   2            By:
RANGE:  0-50 PCT             Asgn'd:              Date:   3 / 27 / 97
Remarks:  Term IN I.A. CAB
```

☑ Process Location Agrees with Loop Sheet Service Description
☑ Service Description Agrees with Console
☑ Process Connection Complete and Correct
☐ **Transmitter 5-Point Check:**

Transmitted Signal	Console Reading
4.0 mA	0.03
8.0 mA	13.3
12.0 mA	26.2
16.0 mA	49.1
20.0 mA	52.5

☐ Loose Wire Alarm Enabled and Verified
☐ Reconnect Verified
☐ New Xmtr Model: _____ S / N: _____
☐ **Valve Check:**

Console Value	Valve Position	
0%	_____	☐ 5% Check
50%	_____	☐ 95% Check
100%	_____	

☐ Valve Fail Position   Open   Close
☐ New Valve Model: _____ S / N: _____

☐ Switch Wired Correctly: No   NC
☐ Alarm on Open Ckt / Information on Close Ckt
☑ Solenoid Activated Deluge @ 40%
☑ Junction Box Grounding
☑ Junction Box Terminals
☑ Junction Box Terminals Sealed
☑ Sign-off Completed

Printed: 06-May-97
Page: 1

Figure 3-19. Example of an inspector checklist (partially completed).

- SAT test results
- SAT certificate
- Validation test results

- Availability and performance guarantee tests
- Additional certificates, where applicable

FAT Documents

For major equipment purchases, most companies require a FAT. A FAT is conducted to determine and document that the equipment and software operate according to the specifications and includes functional, fault management, communications, support systems, and interface requirements. The FAT may be conducted at the manufacturer's facility, the main automation contractor (MAC) or system integrator site, or a third-party location such as a panel shop.

With the PAS, the equipment is often set up in what is referred to as a *staging area*. This is where all purchased equipment and software are demonstrated to prove they meet the aforementioned criteria. If the MAC is responsible for supplying the entire PAS,[2] then the system will include the main control system—usually a DCS, BPCS, or SIS—as well as other systems (e.g., TPPS and interfaces to the enterprise network) that will be interfaced to the PAS.

Criteria to be met are typically an agreed-on combination of vendor claims and performance guarantees and the purchaser's specifications. The specifications are part of the FAT plan described in the following text. The agreement is part of the purchase order for the equipment or system(s).

Usually, people from multiple groups (process, project, and controls engineers; Operations personnel; I&E technicians; and CSTs) meet with the vendor to witness the FAT.

For a successful FAT, it is important to have a good FAT plan. The procedures to be followed should be well defined and documented. Tests must be carried out in the correct sequence because the results of one may affect another.

The plan should include the following:

- The steps to take during the FAT
- A list of equipment and items to be tested (e.g., all loops, only inputs, and only interfaces to other systems) and the functionalities to check

---

2 PAS refers to the entire process automation system. It includes basic, advanced, and safety control systems as well as interfaces to other systems such as the business network and third-party systems.

- Applicable performance criteria

- Personnel responsible for performing the test(s), making corrections, and paying for changes, for example

- Procedures for handling errors or system failures (e.g., hardware or software) and how these should be addressed and documented, along with criteria by which the completion of a test is determined

- Test methods to be used

- Pass/fail criteria

- Test documentation forms and procedures

After a successful FAT, the MAC, in this case, usually issues a certificate to the customer, and then the system is authorized for shipment and shipped to the customer's plant site.

During the FAT, the CST and other team members can examine the equipment in operational mode before it ships. If errors or problems are found, they can be resolved before the system is shipped to the plant, resulting in a less problematic start-up. When the group that participates in the FAT is the same group that is involved in the start-up, the group members have a "head start" on working with this system and with each other before the SAT, discussed later in this chapter, and start-up. Being involved with a properly conducted FAT is one of the best ways for a CST to learn about PAS equipment and how to work with others to prepare for a successful start-up.

SIT Documents

Because the PAS is a highly integrated system and it interfaces to other systems (see Figure 3-15, network topology diagram), many companies require that SITs be performed. SITs are used to test the communications and data flows between systems and through interfaces. Because this testing has to occur at the site, especially to interface with the plant business network, the term SIT contains the word *site* as in *site integration testing*. As with the FAT, a good SIT plan should include the following:

- Personnel involved with the testing

- Test methodology

- Equipment and software to be tested

Typically, information technology and/or software architects who were involved during the FAT, participate in the SIT.

SAT Documents

Many companies now require that a SAT be performed. The SAT occurs after all the equipment and software are installed at the plant site. It is conducted to determine and then document whether or not the equipment and software still function as they did at the factory. In addition, during the SAT, the equipment is checked as it is connected to final control elements, such as valves. These documents and procedures may be referred to as *functional tests* which include *cause and effect* (interlock) and logic checks.

As with the FAT, a multidisciplinary team is involved. First, a SAT test plan is written, usually by a multidisciplinary team but, if possible, a team that includes people other than those who performed the testing to avoid bias. Similar to the FAT plan, the SAT plan should include, at minimum, the test methods used, the functionalities to be tested, the pass/fail criteria, and the appropriate test documentation to use. Once testing is completed, satisfactorily completed tests are signed off on and project completion is tracked. Unsatisfactory tests are added to a punch list and equipment or control logic fixed or updated and retested at a future date.

After it has been determined that the equipment and software meet the criteria (pass/fail) and that the SAT is complete, the tested system is allowed to become operational.

Relevant Standard

International Electrotechnical Commission

- IEC 62381, Ed. 2.0, *Automation Systems in the Process Industry – Factory Acceptance Test (FAT), Site Acceptance Test (SAT), and Site Integration Test (SIT)*, February 1, 2012

Process Validation Documents

According to 21 CFR 820, *Quality System Regulation*, process validation is defined as follows:

- **Process Validation** – "[E]stablishing by objective evidence that a process consistently produces a result or product meeting its predetermined specifications."

As with FAT, SIT, and SAT, objective evidence is established by setting up criteria before testing to prove the results consistently provide acceptable output and to document this evidence along with results and conclusions. Validation improves processes and thereby ensures consistent, high-quality products, product safety, and cost savings.

The FDA inspects product manufacturers (e.g., US pharmaceutical plants and plants that provide products for human consumption) to verify that they comply with good automated manufacturing practices. Validation and the resulting documentation set up a baseline for these types of plants or the parts of these plants being started up. For example, many parameters (e.g., alarm set points and ranges) used by the PAS are recorded. This is the starting point (or *baseline*) for the set of information that is documented when an inspection occurs. If any changes are made to a plant of this type, it must be revalidated and new settings for the parameters must be recorded.

Good documentation practices (GDPs) should be used when recording data during validation. Examples include the following:

- Use black indelible ink.

- Do not use correction fluid.

- Cross out changes with a single line, and initial and date the change and all other entries (see Figure 3-20).

- Do not use ditto marks or arrows.

- Fill in all spaces.

- Mark unused spaces with N/A or a diagonal line through the box on the form.

- Where appropriate, the reason for these alterations must be noted (e.g., "E.E." for entry error).

As a CST, you may work with validation professionals (see Chapter 4) using calibrated meters to test field instruments. You may also work with them on a PAS used in the manufacture of pharmaceutical products and other systems subject to formal process validation. The software used in these manufacturing processes must also be validated.

Tag	EU	Range	Low Alarm Limit	High Alarm Limit	Initials	Date
TIC3501	Deg.F	50-150 °F	55 °F	145 °F	DRB	6/23/2001
FT3602	SCFM	10-350 SCFM	15 SCFM	345 SCFM	DRB	6/23/2001
FT3603	SCFM	10-350 SCFM	15 SCFM	345 SCFM	DRB	6/23/2001
PIC3402	PSIG	0-100 PSIG	N/A	N/A	DRB	6/23/2001
PT3402	PSIG	0-100 PSIG	5 PSIG	~~100 PSIG~~ 95 EE DRB 6/23/2001	DRB	6/23/2001

Figure 3-20. Example of corrections using GDP.

You may also be asked to serve as a witness for validation testing either before or after the software or equipment (or both) is delivered to the plant site, during FAT, SIT, or SAT.

Validation records must be retained for several years and handled by special methods. It is important for CSTs to learn this methodology if they will be involved with this work.

Relevant Regulation

US Food and Drug Administration

- 21 CFR 820, *Quality System Regulation*, April 2010

3.5 Documents Used by the CST

Each contractor or company has its own procedures and common practices for instrument loop checking and functional testing. In the author's experience, a loop folder is the best way to perform these functions because it contains all the necessary documents for a successful plant start-up.

Plants have different methods for performing loop checks. Loop folders are often used as part of the methodology. One of the jobs the quality inspection team typically performs is creating the loop folders. If it is not completed by this team, a set of loop folders may be a deliverable by the engineering company responsible for the project. A loop folder is usually a physical folder (but it can be electronic) that contains an assortment of documents related to the instrumentation for a single process control loop (i.e., there is one loop folder for each control loop that will be loop checked). The contents of a loop folder are defined in a project's standards and requirements, but they typically include the documents checked off in Table 3-2. The quality/inspection checklist (see Figure 3-19) and usually all other items checked off in the Loop Folder column of Table 3-2, are included in the loop folder that the CST uses when loop checking and, if necessary, during recalibration. Table 3-2 also indicates the documents the CST is most likely to use during the project and start-up (CST Redlines column).

Redlines are performed on the documents in this folder as necessary to indicate mistakes, updates, and/or changes. See Chapter 5 for a discussion of drawings and redlining. The person who performs the redlining is responsible for ensuring that

Table 3-2. Typical documents required for start-up.

Documents	Loop Folder	CST Redlines
Project Gantt charts		
Functional specifications		
Process flow diagrams (PFDs)		
Piping and instrumentation drawings (P&IDs)	✓	✓
General arrangement (GA) drawings and plot (location) plans		
Instrument specification sheets	✓	✓
Loop diagrams (loop sheets)	✓	✓
Loop check log sheets	✓	✓
Calibration data sheets		
Installation details	✓	
Manufacturers' information documents		
Electrical wiring diagrams	✓	✓
Network topology diagrams		
Logic diagrams and other control logic documentation	✓	✓
Flowcharts		
Standard operating procedures (SOPs)		
Quality/inspection checklist (QA/QC forms) and punch list	✓	
Commissioning activities list (function checks)		
Commissioning checklist and start-up plan		
Factory acceptance test (FAT) plan		
FAT test results		
Site integration test (SIT) plan		
SIT test results		
Site acceptance test (SAT) plan		
SAT test results		
Validation test results		
Availability and performance guarantee tests		

every change made on documents in Table 3-2, CST Redlines column, are updated by a CAD operator (drafter or designer) or engineer when the change is completed.

Note that the CST might also need the documentation that is not checked off in Table 3-2, particularly if troubleshooting must be done or more information is needed. This table serves as a guide for the most common documentation the CST will use during the project.

Often, the P&ID is the main drawing that many of the other drawings are based on and refer to.

Calibration data sheets, installation details, manufacturers' information documents, and instrument spec sheets are typically used before start-up, but as the CST, you may need to access them if difficulties occur during or after the start-up.

Although you may not use all the documents discussed in this chapter, many of them will have an impact on start-up procedures. They might also be discussed at meetings that you attend, such as the project review meetings described in Chapter 4. Therefore, it is important that you know something about the other documents; they will help you realize that other factors have an impact on your job and on plant start-up.

GA drawings and plot plans are helpful during installation, calibration, loop checking, and quality inspection, particularly if the plant (site) is large. These drawings help personnel locate equipment, thus saving time and avoiding confusion.

Loop diagrams help during installation, loop and function check, and the quality inspection because they indicate where a loop is terminated and show the loops a loop may be connected to (e.g., in a *cascade control scheme*).

Loop check log sheets and function check procedures are primarily used during loop and function checks and when complete they go into the loop folder. There should also be dated signatures on these documents to indicate that the process was done successfully or comments that the loop requires further attention.

Calibration data sheets are primarily used during calibration; however, when there are problems during loop or function checks, they may be reexamined and redlined, and eventually the latest version ends up in the loop folder. When recalibration occurs, some facilities find it helpful to record the original readings along with the new readings so a history is maintained. Calibrations generally include *as-found* and *as-left* data and must include engineering units (EU). An example of an EU is standard cubic feet per minute (SCFM).

Installation details are "typical" for each type of instrument (e.g., Coriolis flowmeter) and are used during instrument installation and later during quality inspection, as needed. These may or may not be in the loop folder, depending on the project requirements.

An instrument "spec" sheet (also called *data sheet*) is used during calibration, loop check, and quality inspection. There is one spec sheet for each instrument with the instrument designation, often referred to as *tag name*, at the top. Each spec sheet will go into its respective loop folder.

Electrical wiring diagrams are used during installation, loop check, and the quality inspection. They too are often included in the loop folder, particularly for digital equipment such as motors. The electrical diagram most often found in a loop folder is the electrical "one-line," typically indicating connections between field equipment (motor) and the motor control center.

Loop folders may be reused for reference after start-up, but the updated electronic versions of the files that are in the Table 3-2 CST Redlines column are the latest version. The hard-copy loop folders are kept as a history of the project.

The plant start-up team, discussed in Chapter 4, should be provided with a complete set of PFDs, P&IDs, loop diagrams, log sheets, electrical wiring diagrams, and installation details. The PM and department heads will determine what other documents are necessary. The team should have access to the documents well in advance of commissioning. Getting documents early should give the CST ample time to study the information during idle work periods so he or she is prepared to perform the necessary project-related activities. Because numerous revisions will be made, and updates will be distributed to reflect changes occurring between day one of design and day one of start-up, each person (including the CST) must keep his or her copies up to date and properly dispose of outdated copies.

One set of loop check log sheets should be maintained. They are usually kept at the operator's console to be marked as each loop is checked. The completed original is then typically placed in the loop folder associated with that loop. If there is more than one loop on the loop check log sheet, some organizations highlight the loop on the page, create multiple copies of the loop check log sheet, and place the copy highlighted for the completed loop in its individual loop folder.

You should have copies of loop sheets and electrical drawings to use while on the plant floor or in I/O rooms so you may communicate effectively with the operator at the console.

Manufacturers' documents and SOPs for start-up must also be distributed to all necessary personnel, and may be sent to the Engineering, Maintenance, Operations, and Training departments.

As changes occur during start-up, they must be made to the drawings, thereby ensuring that as-built drawings are being created. The procedures for making these changes are described in Section 3.6. The PM maintains the project Gantt chart and keeps the start-up on schedule and within budget.

Relevant Standard

International Electrotechnical Commission

- IEC 61131-3, Ed. 3.0, *Programmable Controllers – Part 3: Programming Languages*, February 1, 2013

3.6 Control of Project Documents

At many plants, the Engineering department maintains the originals of documents and drawings. This department often has a group working with (or for) it, designated as *Document Control*. During a project (engineering, construction, and start-up phases), Document Control personnel are responsible for maintaining up-to-date documents by ensuring that proper procedures are used to redline changes on originals and that these documents are updated and added to the EDMS with the proper revision information and approvals (signatures).

Other plant departments are responsible for maintaining their copies, which may or may not be up to date. Document maintenance must be checked periodically (with Document Control's version), especially if one has made a copy (paper or electronic).

For handy reference, loop sheets and electrical wiring diagrams are typically kept in I/O rooms and in other cabinets and enclosures. Again, this can be a problem because you do not know if you are looking at the latest version. Because the practice of maintaining documents and drawings is dictated by plant policy, it leads naturally to the topic of MOC, which is covered more fully in Chapter 5.

Summary

Many drawings and documents are used during the construction of a new plant and during the start-up phase of a project. These drawings and documents come in many formats and are located in different areas of the plant. It is easier for someone at the plant to gain access to them if he or she has a computer and these drawings and documents are on the network. As a CST, however, you will still need to have paper copies when working in the field and in the shop on equipment related to start-up.

As start-up commences and equipment problems occur, an understanding of the different drawings and related documents will help you when installing and troubleshooting new equipment in the plant. Your understanding of the drawings will help you quickly obtain the proper documents to complete your job and keep the start-up on schedule.

Review

3.1 What are the differences between a piping and instrumentation drawing (P&ID) and an installation detail?

3.2 In the past, what material was used for P&IDs and other engineering drawings?

3.3 What is the advantage of scanning documents?

3.4 Which document is used to track project activities and status and to coordinate between various activities?

3.5 What does SOP mean, and why are SOPs important?

3.6 Name some of the standards used when drawing documents.

3.7 What are the disadvantages of scanning documents?

3.8 What are the "layers" used for on computer-based P&IDs?

3.9 What does EDMS stand for?

3.10 Why is equipment manufacturer information important? What can these documents tell us?

3.11 What are "bubbles" used for on engineering drawings?

3.12 Name some locations where documents are normally found.

3.13 What is an energy balance?

3.14 What are the differences between pneumatic, electrical, hydraulic, electromagnetic, and logical connections? How are these shown on P&IDs?

3.15 What types of controllers use Ladder Logic?

3.16 What is a Gantt chart? How does the CST contribute to the Gantt chart?

Recommended Reading

Keong, Kwek. "Project Network Plan – An Effective Project Management Tool for Planning and Control." White paper presented at ISA EXPO 2006, Houston, TX, October 2006. https://www.isa.org/store project-network-plan-an-effective-project-management-tool-for-planning-and-control-isa-expo-2006/120671.

McAvinew, Thomas, and Raymond Mulley. *Control System Documentation: Applying Symbols and Identification*. 2nd ed. Research Triangle Park, NC: ISA (International Society of Automation), 2004.

Meier, Frederick A., and Clifford A. Meier. *Instrumentation and Control Systems Documentation*. 2nd ed. Research Triangle Park, NC: ISA (International Society of Automation), 2011.

4
Working with Others

Start-Up Team Organization
CST Interaction with Other Personnel
Chain of Command

To work effectively during a start-up, the CST must understand the related equipment, control, and process aspects of the plant. An understanding that management, engineering, programming, mechanics, instrumentation and electrical systems, plant operations, chemistry, safety, and environmental issues and regulations all play a part in the project is key. Most importantly, the CST must have the ability to work with different people. Crucial to a successful start-up is the ability to effectively communicate and work with all the disciplines, groups, and people involved in the project.

This chapter will discuss the departments, groups, and disciplines involved with the project, the duties of each, how these duties affect both the project phases and the CST's involvement with start-up and commissioning, and how as a CST you will interact with team members during the project.

Note that the plant and corporate organization described in this chapter may not correspond exactly to your facility, but the basic principles hold true. Many plants now have a single electrical/instrument engineer (i.e., engineering and/or technical representative; see Figure 4-1), often the result of corporate downsizing, and not all the departments and personnel described in this chapter may be found in your plant.

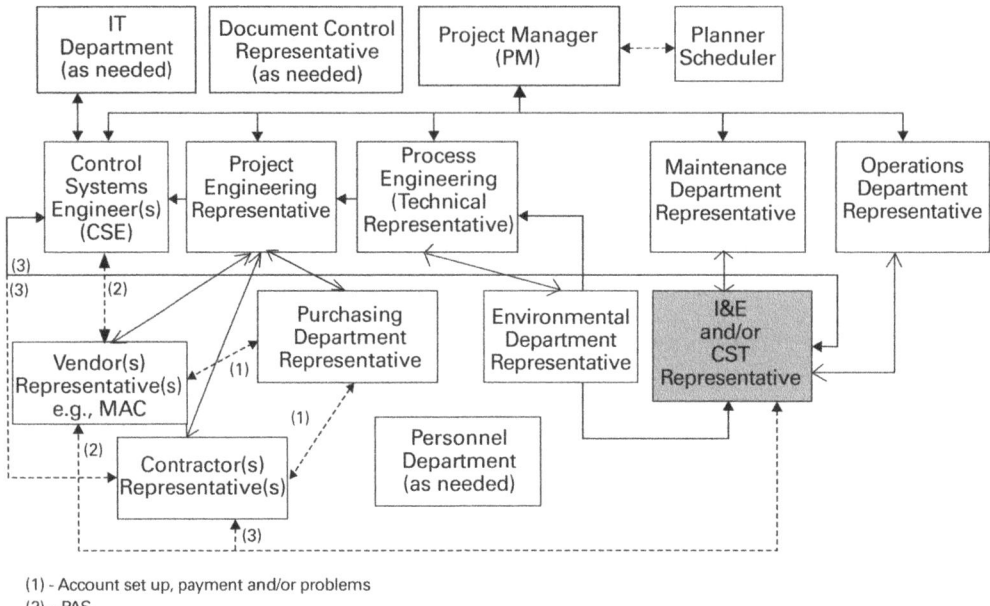

(1) – Account set up, payment and/or problems
(2) – PAS
(3) – Instrumentation, for example
Note: Contractor(s) representative(s) include quality/inspection representative and/or other construction personnel.

Figure 4-1. Typical project (start-up) organizational chart.

It is also common for a CST to have to take on more responsibility than is described here. In addition, not all members of the project team may participate in the start-up. However, during start-up a CST may need to know where to go and whom to contact if there are problems or questions pertaining to project design.

Typically, at a project kickoff meeting the project manager (PM) will announce the start of the project and invite leaders or managers from each department or group to participate as a way of representing all involved parties, including those from outside the plant. During this meeting, an organizational chart showing the project team members' names and disciplines and the reporting structure will be distributed and discussed. During the formation of the project team, members of the Instrument and Electrical (I&E) department(s) are typically included. For large projects, a construction management company may be in charge until the construction is complete (known as *mechanical completion*), after which this responsibility is turned over to the owner/client company. An architect or engineering firm is also part of the project team. These entities are responsible for subcontracting most of the work and bidding out/purchasing any equipment other than that specified as "owner furnished." These types of companies, depending on their scope of work, are often referred to as *engineering, procurement,* and *construction* (EPC) *firms* and may be represented under the items denoted in Figure 4-1 as Project Engineering and/or Process Engineering Representatives.

Figure 4-1 illustrates a typical organizational chart showing representatives from each of the disciplines, departments, and groups involved in a project. Although they are not commonly used in organization charts, arrows have been included in this example to show the interactions between representatives. The CST block is highlighted. There may be one or more representatives from each department on the team. Each of the items in Figure 4-1 will be discussed in greater detail below.

During the kickoff meeting, one of the first orders of business is communicating each team member's role and how much of his or her time will be required. The kick-off meeting also establishes how often the team will meet. The team distributes, discusses, and updates a checklist or Gantt chart (see Chapter 3) of all project and/or start-up activities at each meeting.

Although it may not be done at the kickoff meeting, it is important to decide early in the project whether all the work is to be completed before start-up or whether areas of the plant will be started up in phases per a start-up plan.

Meeting frequency varies depending on the stage of the project. For example, the team may meet once a month after the kickoff meeting during the project design phase. As the start-up date gets nearer (during plant construction), more meetings will be held to ensure that all work efforts are coordinated and that the start-up will be safe and successful.

As a CST, you will probably be a member of the start-up team as part of the I&E department, and you will most likely work with people from the Operations, Control Systems, Project Engineering, and Process Engineering departments daily. You probably will also be involved with contractors and vendors. It is therefore important for you to know who these people are and what their job functions are during the start-up so that you may interact with them efficiently. Figure 4-2 shows one plant's project team assignment form.

4.1 Start-Up Team Organization

4.1.1 Operations Department

Operations, also known as *Production* or *Manufacturing*, includes the plant operators and their direct supervisors, who are usually foremen or area supervisors. Operations personnel may sit at the operator console of a process automation system (PAS), also known as a *human-machine interface* (HMI), or work in the field. The operator's basic job is to run the plant by ensuring that process control points (e.g., temperature, level, and

Project Team Assignment

PN: _____ Project Title: _____

Meeting 1: Scope

Date:	Time:	Location:

Team Members

Category/Department Representative	Responsible Supervisor	Assigned Person
Hourly Maintenance		
Hourly Operations		
Process Engineer		
DCS Engineer		
Environmental		
Safety		
Production		
Maintenance		
Purchasing		
Project Engineer		
Maintenance Engineer		

Project Description

_____
_____
_____
_____
_____
_____
_____
_____
_____
_____
_____
_____

Figure 4-2. Project team assignment form.

pressure) are maintained and that feed and product materials flow properly to make the product according to the specification.

An operator who perceives a problem with the process may call the Maintenance department to help resolve the problem. If it is determined that it is a process control problem, the CST and any other required I&E technicians must work with the operator to determine what the problem is and to resolve it.

As a CST, during start-up you will interact with Operations more than with personnel from any other department. You must do this for safety, check-out, and troubleshooting, as well as for routine operational and process purposes.

During loop check, the plant operator is involved with *stroking* (opening to various positions and closing) valves; *bumping* (starting and stopping) motors for pumps, agitators, and other rotating equipment; watching the indicators of these and other types of equipment; and then communicating the results to the CST. The CST is usually not in the control room with the operator during these actions, so you cannot see what the indications on the PAS are. You may be in the field watching the valve or motor, or you may be in the I/O (input/output) room feeding or reading a signal (e.g., 4–20 mA) to or from the field. You may also be working on field devices, performing actions such as replacing current-to-pneumatic converters or transducers, reversing the action of a valve, or changing the calibration of a transmitter. You may also be in the I/O room replacing a printed circuit board or checking terminations from the field to the PAS or junction boxes. Whatever job you are performing as a CST, you will be in communication (often via a two-way radio) with the operator so you can receive feedback (PAS indication) regarding the system the operator is monitoring. As discussed in Chapter 2, the two of you must be in close communication to ensure safety and efficiency.

During check-out, the operator can also answer many of your questions, for example, whether a new valve appears to be functioning properly when it is installed or whether a new flowmeter reading makes sense. Conversely, the operator may call on you to help solve a problem. It is essential that you and the operator communicate with each other clearly.

Troubleshooting (which is often associated with loop and function checks and done when there are problems, process changes, or plant additions) requires a similar level of communication between you and the operator.

4.1.2 Control Systems Engineer

A control systems engineer (CSE) may be assigned to the plant Project or Process Engineering, Maintenance, or Operations departments. In any event, a CST will most likely work closely with a CSE to ensure that communication links between field devices and the PAS are working properly. As a CST, you may even work with the CSE to ensure that communication links with other systems (e.g., enterprise network) and third-party packaged systems (TPPSs) work properly. You will work particularly closely with a CSE if you are involved with PAS configuration (e.g., graphics and loops) and programming (e.g., automation sequences).

The CSE's duties prior to and during start-up are to ensure that all PAS equipment has been received and installed, all software has been configured or programmed, and loop and function checking is ready to commence. Often the CSE, CST, and Operations and Maintenance personnel work together to "check out" the loops for the part(s) of the plant ready to be started up. You will work with the CSE and with Operations and Maintenance personnel if there are problems with the equipment (hardware) or software, or if changes must be made. If you are involved with configuration and programming changes (which may include the I/O database, graphics, automation sequences, and historian [data collector]), then you should ensure that the software is backed up by you, by the CSE, or by other people responsible for these activities before programming changes are made.

4.1.3 Project Engineering Department

The Project Engineering department will have been involved in the design of the plant and purchase of equipment for the start-up of a new or modified plant. They will have interacted with engineering firms (if used), vendors, contractors, and the maintenance group (including I&E). They are therefore knowledgeable about much of the equipment and how it should perform during start-up and later during normal operations. This means they must work with the CST to get the plant online, and the CST will need their help if there are any questions or concerns regarding instruments, equipment, plant design, or vendors.

During the construction phase of the project, the duty of the project engineer, who may also be the PM, is to make sure that equipment and instrumentation that were ordered for the plant have arrived on-site, to make sure they are installed according to the design and drawings, and to arrange for manufacturers' representatives or technicians to be on-site at the appropriate time. The project engineer must ensure that personnel and materials are on schedule and within the project budget. The CST will work with the project engineer if there are problems, or if he or she has questions about this equipment.

Document Control personnel may be part of the Project Engineering or Information Technology (IT) department. They may even have their own department. Regardless, project and process engineers and most other disciplines work with the Document Control group to maintain and get up-to-date documentation and drawings for start-up.

4.1.4 Process Engineering (Technical Representative)

Process engineers will also have been involved in the design of the plant. They work with the Project Engineering department to ensure that the proper equipment and controls are purchased to make the product, and they most likely will also have worked with vendors during the design stage to decide which equipment will work best. Their role includes developing the mass and energy balances associated with the desired product(s) the plant will produce.

The difference between this group and the Project Engineering group is that process engineers usually have chemical engineering degrees, whereas project engineers usually have mechanical or electrical engineering degrees and additional (but optional) Project Management Professional certification. The CST would therefore work with a process engineer on process control questions or concerns (e.g., transmitter ranges) but with a project engineer if equipment design issues come up (such as the need to change a pump to provide more flow and your subsequent involvement with transmitter re-ranging). Increasingly companies are merging the roles of engineering and technical specialists in corporate departments. The aging workforce and the lack of growth in the number of engineers and technical specialists graduating each year are forcing this action. As a result, people can become very busy during start-up. Therefore, it is important for a CST and others to prioritize their work.

A process engineer's duties during start-up include monitoring process control points (temperature, pressure, level, flow rate, etc.), product quality, and product quantity. The process engineer may also be the one who sets the control points (e.g., set points and specifications) and establishes recipes for the different products or grades that the process is to produce. As mentioned, process engineers have often been involved in the design of the plant, which can make them good contacts for information pertaining to instrument calibration and loop configuration for process control.

Validation is the process of checking to ensure that a product, service, or system meets specifications and fulfills its intended purpose. These are critical components of a quality management system such as ISO 9000. Process engineers may have to function as part of a site acceptance team or they may be full-time validation professionals, particularly if the site requires US Food and Drug Administration (FDA) validation of the process. Some companies hire a company or contractors to perform the initial validation (baseline) of a new process prior to start-up, and afterward site personnel follow the Management of Change (MOC) procedure to revalidate the new process as changes occur. Because procedures for validation of a manufacturing process include instrumentation and PAS, as a CST you may be involved with these activities. See Chapter 5 for more information about validation.

Relevant Guidelines and Requirements

American Society for Testing and Materials

- ASTM E2537-16, *Standard Guide for Application of Continuous Process Verification to Pharmaceutical and Biopharmaceutical Manufacturing*

International Standards Organization

- ISO 10006:2017, *Guidelines for Quality Management in Projects*
- ISO 9000:2015, *Quality Management Systems – Fundamentals and Vocabulary*
- ISO 9001:2015, *Quality Management Systems – Requirements*

US Food and Drug Administration

- *General Principles of Software Validation; Final Guidance for Industry and FDA Staff*, January 11, 2002

4.1.5 The Project Manager and Other Management Personnel

A PM typically heads up the project and start-up teams. The PM's basic duty during a project is coordination—bringing together all project activities from inception and design through purchasing, procurement, and receipt of materials for commissioning and start-up. The PM will report on project status, including costs, start-up timing, and management and production issues; keep track of costs and the schedule; and ensure that adequate resources (material and personnel) are available to support the project. The PM may also work with a planner/scheduler (see Figure 4-1) to develop and update the project schedule (Gantt chart) or, depending on the size of the project or plant organization, do this work himself or herself. It is important that a CST stays well informed of the project schedule and start-up activities through information provided by the PM.

The PM may also be responsible for selecting the project and start-up teams, which may be the reason a Certified Control Systems Technician (CCST) would be picked for a given project over noncertified CST and I&E technicians. Whatever your certification level, if you have been picked to serve on a team, you will report your work status to the PM at each project review meeting and via written communication, such as progress reports. It is important for you to arrive promptly at each project review meeting, be prepared to give a progress report on your area, and answer questions or voice your concerns.

Managers from other departments (e.g., Operations, Engineering, Maintenance, Purchasing, and Human Resources) will also get involved with the start-up either as the representative from their department, especially if there are only a few people in the department, or as the need arises. For example, if a contractor committed a major safety infraction, the resolution of the problem must be elevated from an operator to the operations manager, who gets the purchasing and/or personnel manager involved, as required.

4.1.6 Maintenance Department

As a CST, you will interact with I&E technicians and other CSTs who will work closely to install equipment and commission and start up the plant. Together, you will most likely calibrate and install instrumentation, run and terminate field wiring, terminate I/O (input/output signals to the PAS), and complete loop checking and function tests. In addition, you will work with pipe fitters, electricians, mechanics, and machinists to make sure the equipment they work on is properly installed, aligned, and lubricated before check-out and start-up commence.

Your basic duties as a CST—calibration, installation and replacement, termination and power-up, loop and function checks, and PAS configuration—may overlap those of an electrician. However, some plants make a clear distinction between what each craft does. These distinctions may be spelled out in the plant's union contract or operating practices. For example, one of the plants surveyed for this book assigns electricians the task of completing field terminations for voltages higher than 24 VDC. However the different tasks are assigned, the PAS electrical requirements guarantee that CSTs will work with personnel from the electrical craft.

The basic duties of a machinist, pipe fitter, welder, or other mechanical craftsperson during start-up are the installation of major and minor pieces of equipment, welding, and the completion of pipe runs. As a CST, you will work with personnel from these crafts when you must install or remove instrumentation during start-up because of changes to the process. For example, you may need to wait for a tap to be installed by a pipe fitter before you can add a new sensing probe.

In one plant surveyed, CSTs are assigned to install tubing for valves only up to a certain diameter; larger ones are handled by a pipe fitter.

4.1.7 Environmental Department

A CST's interactions with this department may be limited, or a CST may be assigned to this department full-time. This depends on the amount of environment-related instrumentation and process control that is required for the plant. Whatever the relationship, there are many important things a CST is involved with in the environmental field, particularly because the CST is responsible for installation and maintenance of the instrumentation providing inputs to systems, such as the PAS, the continuous emission monitoring system (CEMS), and the predictive emission monitoring system (PEMS), described later in this section.

The Environmental department interfaces with government agencies (e.g., US state and federal environmental protection agencies) and is responsible for getting permits for a new facility before start-up. Permits define the nature and maximum allowed quantity of effluents into the air, ground, and water. These effluents are often measured using air and water monitors that were designed by the process engineers and purchased by the project engineers. Generally, ground disposal is not a continuous process; therefore, instrumentation, except for laboratory analysis, is often not required. As a CST, you may or may not be involved with calibrating these monitoring instruments and working with any connections to associated systems, but you will definitely be involved with loop checking them. You may also be involved with checking systems against standards, that is, using a known measurement from a standard sample to verify that the instrument is calibrated properly. If these critical environmental monitoring and control systems do not function properly or are showing, for example, out-of-range readings, you should notify the appropriate personnel.

Some of the duties of an environmental specialist during start-up include ensuring that permitted stacks are tested, standards are available for calibrating environmental monitoring instruments, continuous monitoring is in effect, and data is being gathered for the Environmental department. You may be asked to assist the environmental specialist. Some data must be printed and kept for relevant government entities. If these data come from a chart recorder or other method of printing (hard copy), you may be asked to obtain this documentation daily or maintain it if problems occur. If the data come from the PAS, you may be asked to set up the reports for the Environmental department by doing the necessary configuration.

The US Environmental Protection Agency (EPA) Environmental Technology Verification (ETV) program verifies the performance of innovative technologies that have the potential to improve protection of human health and the environment. Verified technologies apply to all environmental media—air, water, and ground—and to work, they require instrumentation and controls that a CST is responsible for.

The many verified technology systems a CST may work on include the following:

- Ambient ammonia sensors
- Ambient fine particulate monitors
- Ammonia CEMSs
- Multi-parameter water monitors or water quality probes
- Portable water analyzers/test kits
- Turbidimeters
- Groundwater sampling devices
- Wellhead monitoring technologies
- Volatile organic compounds emission control technologies
- Flowmeters for various applications

Relevant Internet Reference

"Environmental Technology Verification (ETV) Program Site Characterization and Monitoring Technologies Center," US Environmental Protection Agency (EPA), https://semspub.epa.gov/work/HQ/189908.pdf

Continuous Emission Monitoring Systems

Continuous emission monitoring systems (CEMSs) fall under the EPA's ETV program.

The CST may work with the instrumentation and personnel involved with the CEMS used to monitor combustion systems (incinerators, power plants, refineries, etc.), making sure emissions remain within legal limits. CEMSs provide real-time emissions data including the following pollutants: nitrogen oxides (NOx), carbon monoxide (CO), and sulfur dioxide (SO_2), along with oxygen (O_2) and various operating parameters such as temperature and opacity. These are monitored on a continual basis.

CEMSs are required under some EPA regulations for either continual compliance determination or determination of exceedances of the standards. Performance specifications are used for evaluating the acceptability of the CEMS at the time of or soon after installation and whenever specified in the regulations. These evaluations are conducted during a site acceptance test (SAT) and periodically after plant start-up, as required, and may involve the CST.

Predictive Emission Monitoring Systems

Predictive emission monitoring systems (PEMSs) are used for monitoring pollutant emissions without traditional hardware analyzers by using process parameters and their known relationship to pollutant concentrations. Correlation tests are performed and the results are compared to operating data to estimate the amount of emissions from the source. Because sensors are still needed, the CST is involved with start-up and maintenance of these sensors and with initial testing of the PEMS.

Leak Detection and Repair Programs

US refineries are required to implement leak detection and repair (LDAR) programs for processes and streams described in the National Emission Standards for Hazardous Air Pollutants from Petroleum Refineries, known commonly as the refinery maximum achievable control technology rule. Because valves and connectors can cause chemical leakage, the CST and the I&E department would be affected by this rule.

> **Relevant Technical Report, Recommended Practice, Standards, and Act**
>
> <u>International Society of Automation</u>
> - ISA-TR52.00.01-2006, *Recommended Environments for Standards Laboratories*
>
> <u>US Federal Register</u>
> - 40 CFR 60, *Standards of Performance for New Stationary Sources*, July 2017
> - 40 CFR 63, *National Emission Standards for Hazardous Air Pollutants for Source Categories*, July 2017
> - 42 U.S.C. §13101 et seq. (1990), *Pollution Prevention Act*, Environmental Protection Agency (EPA)
> - *Smart Leak Detection and Repair (LDAR) for Control of Fugitive Emissions*, June 1, 2004, American Petroleum Institute (API)

4.1.8 Purchasing Department

A CST's interactions with the Purchasing department may be limited, but it is worth mentioning that this department is also part of the project and start-up teams.

The basic duties of a purchasing agent or buyer before start-up are to aid in bidding or getting a good price for a necessary item or service; purchasing the item or service; and arranging shipment, delivery, receipt, and payment. During start-up, this same agent will continue to buy items as necessary and arrange for vendors and vendor representatives ("reps") to come to the plant if problems or questions arise. As part of your role as a CST, you may need to take a purchase requisition to the Purchasing department so a purchase order (PO) can be generated for a desired item or service. You must know the procedures for doing this and whom to work with.

You may become involved with purchasing if a new instrument or group of instruments fails to function. The Purchasing department must then notify the manufacturer and possibly withhold payment or work with you to request replacement.

If changes to the process require that additional instrumentation or equipment you are responsible for be purchased, you may be required to track these items from order entry to receipt. If this is the case, it is important for you to know how to track the PO.

As a CST, you also may need to help the Warehouse or Receiving department receive items if they do not have the necessary expertise to identify the equipment. You do this by physically inspecting the received item and by cross-checking against a PO.

4.1.9 Human Resources (Personnel)

The Human Resources (HR) department, also known as *Personnel*, comes into play during start-up if safety, training, and manpower issues arise. Staff in this department will have been assigned to each of these areas. Safety was discussed in detail in Chapter 2, so we will not discuss it here except to reiterate that it is one of the top priorities during start-up.

The record-keeping function associated with safety may be performed by the safety supervisor or safety engineer in the HR department.

Although most training for a new plant occurs before start-up, there is still a learning curve. New things are learned by making mistakes, gaining input from representatives from Operations, and implementing changes. Because of this, Training department personnel are often present during start-up, conducting training, making changes to written procedures, and learning about the equipment. Start-ups often require overtime, so any payroll or grievance problems that arise may need to be resolved by the HR department.

A CST will interface with the HR department for safety training; safety concerns; investigation of accidents or near misses; CST and instrument training; maintenance procedures; and administrative issues such as pay, staffing, vacation, and attendance. If your company encourages certification in the CCST program, the HR department may aid you by coordinating and arranging this training, allowing time off to attend classes or study, and even reimbursing you for certain expenses. Your desire to achieve this goal will be rewarded monetarily, through job status, and through enhanced self-esteem.

4.1.10 IT Department

Often the enterprise network and PAS will interface; therefore, the personnel responsible for these systems, IT and CSE, respectively, will interface as well. These interactions occur for the following reasons:

- **Shared network infrastructure** – This include the fiber-optic network and associated equipment such as hubs and routers.

- **Communications between systems** – Some of these communications are bidirectional, whereas others are one-directional (i.e., data is sent from the PAS to the enterprise network, but usually not vice versa).

- **Security and System Maintenance** – Systems must be designed to ensure that security is maintained when the system communicate and during system maintenance (e.g., virus protection, patch deployments, backup, software updates and adding users).

- **Access to documents and drawings** – The documentation described in other chapters of this book must be accessible to members of the start-up team. This information is usually on the enterprise network and accessible (with established security) from network personal computers. To obtain access or if problems occur, the CST must get assistance from IT personnel.

It is important for IT and CSE personnel to work together during the design phase of the project to ensure that the network infrastructure can properly support the interfaced enterprise network and the PAS. If the PAS requires a certain communication network (e.g., Ethernet EEE 802.3X) or type of fiber (e.g., single-mode versus multimode), then IT and CSE personnel must meet, possibly with vendor representatives and engineering design companies, to ensure that the systems will work together when the time for start-up arrives.

Because of the need for these networks to interface, we turn to a short discussion about the international ANSI/ISA-95 standard for integrating enterprise and control systems. The ANSI/ISA-95 models and terminology can be used to determine which information can be exchanged between systems, for example: sales, finance and logistics, and systems for production maintenance and quality. The ANSI/ISA-95 functional hierarchy illustrated in Figure 4-3 shows how the PAS (Levels 1 and

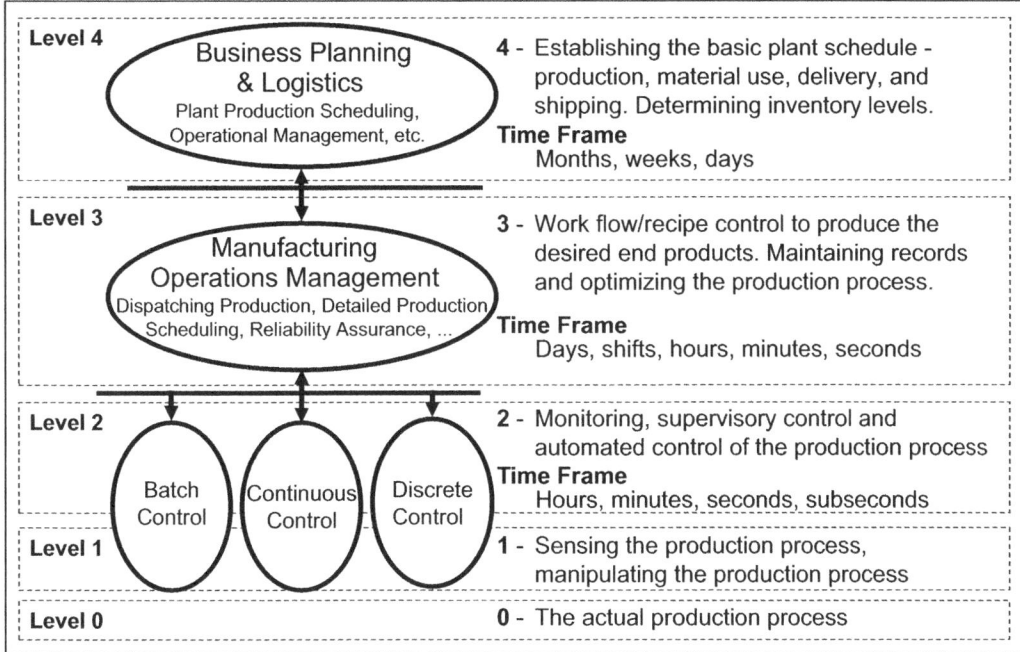

Figure 4-3. Functional hierarchy.
Source: Reproduced with permission from ISA.

2) connects to the manufacturing operations management network (Level 3). Figure 4-3 helps illustrate the interface between production scheduling and operational management (Levels 3 and 4) and automated control of the production process (Level 2). Level 2 indicates the control activities involved in manufacturing, either manual or automated, that keep the process stable or under control.

Relevant Standards and Technical Report

<u>American National Standards Institute/International Society of Automation</u>

- ANSI/ISA-95.00.01-2010 (IEC 62264-1 Mod), *Enterprise-Control System Integration – Part 1: Models and Terminology*
- ANSI/ISA-95.00.02-2010 (IEC 62264-2 Mod), *Enterprise-Control System Integration – Part 2: Object Model Attributes*
- ANSI/ISA-95.00.03-2013 (IEC 62264-3 Mod), *Enterprise-Control System Integration – Part 3: Activity Models of Manufacturing Operations Management*
- ANSI/ISA-95.00.05-2013, *Enterprise-Control System Integration – Part 5: Business-to-Manufacturing Transactions*

- ANSI/ISA-62443-1-1 (99.01.01)-2007, *Security for Industrial Automation and Control Systems – Part 1-1: Terminology, Concepts, and Models*
- ANSI/ISA-62443-2-1 (99.02.01)-2009, *Security for Industrial Automation and Control Systems: Establishing an Industrial Automation and Control Systems Security Program*

International Society of Automation

- ISA-TR88.95.01-2008, *Using ISA-88 and ISA-95 Together*

4.2 CST Interaction with Other Personnel

Meetings prior to start-up should include discussing how the I&E department and the CST will interact with members of other disciplines. CSTs may work in an organization where they are part of a multicraft crew (instrument, electrical, and mechanical). The members of the start-up team should discuss and assign task priorities, establish contacts, and identify the applicable "chain of command," the hierarchy of personnel and reporting structure within an organization.

Because contract personnel, including sales and vendor representatives, may be unfamiliar with plant operation and hazards, plants sometimes assign a key person to serve as liaison between plant personnel and contract personnel (see Chapter 1). Some plants require this person to be fully escorted, which means he or she must always remain with the assigned liaison. The liaison may also help contract personnel procure tools or materials for the job or coordinate work activities with other plant employees. An example is when a contractor needs the liaison to help coordinate with Operations personnel when equipment purchased from the contractor must be tested (or run).

The same holds true for vendor representatives. If a vendor representative is from an electrical or instrument supply house, then a CST may be assigned as liaison. Just before start-up and after it has begun, there may be several other vendor representatives (such as those with TPPSs) who must work with you to get their systems working and checked out.

You must communicate with contract personnel to ensure that what they are doing does not compromise the safety or operation of the plant.

A CST may also serve as the technical lead on projects. One company surveyed uses the job title *Senior Technical Specialist—Automation*. This CST directs the work of the plant technicians, contractors, and vendors. Technical leads are often teamed with a senior engineer who serves as the engineering lead. A CST may also work with vendors, contractors, and plant personnel during factory acceptance tests (FATs) and SATs. See Chapter 3 for more information pertaining to these tests.

4.2.1 Contractors

It is very likely that a CST will work with instrument, electrical, and mechanical contractors before and during the start-up of a new facility. Many projects utilize EPC companies to do the engineering design (architectural, electrical, and controls) and the electrical, instrument, and mechanical installation. As the name implies, an EPC is responsible for the engineering (and design of a project), the procurement of materials (equipment, piping, professional services, and delivery of materials to the site), and full construction (installation and erection of equipment, piping, instrumentation, and electrical). Because mistakes often occur, contract personnel will be in the plant during start-up to correct design and installation errors. Other changes result from process changes during start-up, especially if a completely new plant or process is being started up. It is important that these changes be tracked and documents be updated. These tasks are included in the CST's scope of work.

In addition, electrical contractors will probably work with the CST during loop and function checks, especially if they were responsible for the wiring installation. Mechanical contractors who modify piping will work with you if process changes during start-up require that instrumentation be moved or added. For this reason, a CST must have good communication skills and the necessary expertise to work with a multicraft crew.

4.2.2 TPPS Representatives

Often, plants purchase and install *packaged systems*. These systems may be stand-alone, single-loop controllers, or small control systems that interface to the plant PAS, or at minimum send signals from the system and equipment instrumentation to the PAS. Chapter 3 discussed why these systems often have control systems that are separate from, but interface to, the PAS. As a CST, you may be asked to work with the TPPS representative to help get this new equipment started up.

Some plants show these units on the piping and instrumentation drawings (P&IDs) as "vendor supplied" or "by others," with little detail about their actual operation and control. It is imperative that the plant receive and maintain complete documentation concerning these systems, with drawings indicating tags, logic and controls, and the proper interconnections to other plant systems. If this information is not provided, then after start-up it may be difficult to determine how the system works or to find someone from the original manufacturer to provide support when the system needs repair. Dealing with this missing information can be an expensive and time-consuming business. Where appropriate, the CST should question why this information is missing and press the vendor and project or process engineers to include this information in the plant documentation. Cables connecting the TPPS and PAS and any programming must be completed to allow communication between the two systems before the vendor representative can test connections. See Chapter 3, Figure 3-15, to understand this interaction.

Vendor representatives are normally present during the commissioning and plant start-up phases to assist with the packaged system purchased by the plant owner from the vendor. The representative is familiar with the system's operation because he or she has probably traveled to other sites and started up the same model several times. It is imperative that you learn all you can from the vendor representative about how the packaged system operates and where everything is.

Vendor representatives may know little about the plant they are working in during start-up, so they must work with plant personnel. Typically, vendor representatives work closely with CSTs, particularly if the packaged system is going to interface with the plant's PAS. In this case, you would assist the representative with the commissioning and start-up of this system, its interaction with the plant process and PAS, and the start-up of the rest of the plant.

As a CST, you must understand the impact the packaged system will have on the rest of the plant and know the schedule or sequence of events that is to occur. Things that you work on will be prioritized. For example, the CST must determine if the packaged system must be running before the rest of the plant or vice versa. Chapter 3 discussed how such priorities are reflected in the project Gantt chart or schedule.

The basic duties of a packaged system vendor representative during start-up are to communicate what he or she needs to get the system running (e.g., power, inputs, and outputs). The representative also must communicate any changes that are made to the system and its controls, so that associated drawings and documentation may be updated as necessary. The representative normally works with the operator and CST to complete the commissioning of this system and trains people in the plant by explaining how the system works and how it should be maintained.

4.2.3 PAS Vendors and System Integrators

First, it is helpful to know some terminology. The PAS is referred to in this chapter as an *integrated system* composed of any and/or all of the following:

- Distributed control system (DCS) or basic process control system (BPCS).

- Programmable logic controllers.

- Panel-mounted and field instrumentation ("loops"). Here the term *loop* refers to all instrumentation, equipment, and wiring associated with one control system I/O.

- Connections to a TPPS.

- Connections to systems on other ANSI/ISA-95 levels (e.g., enterprise network); see Figure 4-3.

Main automation contractors (MACs) who have supplied the PAS and other control system(s) (e.g., PLCs, DCSs, or TPPSs) and systems integrators will probably work with you to a great extent during commissioning (loop and function checks) and start-up. This is because all the instrumentation and controls that a CST works with are connected to these systems. As a CST, you can learn from both of these groups. If either group makes changes to loops, databases, automation sequences, or graphics, you may need to ensure that the instrumentation signals match or determine if they need to be retested, the relevant documents that are affected are red-lined, and/or data and software are properly backed up. If your company sends you to train on the PAS, you can work effectively with the MACs and system integrators during commissioning and start-up, especially if they need someone to help with changes to instrumentation and PAS hardware and/or software.

4.2.4 Personnel from Other Plants

Many companies have multiple facilities making the same or similar products. Such companies may design a new plant based on existing facilities and use people with experience from one plant to assist in the start-up of another. After the new plant is running, the company may make comparisons between the new facility and the older one to help determine whether the new one is running properly and is making the quantity and quality of product that it was designed for and expected to make.

Problems that occurred in other plants may have been taken into account by making design changes in the new plant. Meetings or process hazard reviews provide an occasion for anticipating and planning for these problems.

CSTs are often asked to assist with start-ups at "sister" sites, especially if they have experience with a similar process or expertise required at another plant, so you may be asked to travel. In such cases, it is important to prepare for your trip by finding out what materials and tools you might need to bring with you and, if possible, the name of your contact at the other plant.

4.3 Chain of Command

An organizational chart such as that shown in Figure 4-1 serves as a tool for bringing the start-up team together, publishing the team's membership and relationship to others, and helping team members understand the lines of communication and the chain of command. A *chain of command* is a system whereby authority passes down from the top through a series of executive positions or military ranks in which each is accountable to the one directly superior. This system helps define your responsibilities and identify who has the power to fix problems during start-ups when you do not have the authority to fix the problems on your own.

Some groups or departments are not included in Figure 4-1 but must be engaged from time to time to assist with problems. Whether or not they are part of an organizational chart, it is important to know whom to go to get things done and to establish workable relationships with them.

For example, while completing the calibration procedures for a control valve, you determine that it is not working properly. In most cases, as CST you will notify your direct supervisor or the project engineer of the problem. Depending on the project team structure, you might be involved in additional testing and gathering of data for Engineering to specify a replacement. If you are serving as technical lead on the project, you might be tasked with submitting purchase information to the Purchasing department. The engineer or manager will then take steps to procure the valve, such as going through the Purchasing department to order a new one. After the new valve is received, the engineer might pick it up or send you to the Receiving department to pick it up. The engineer will then have the appropriate personnel install the valve after it has been calibrated. During installation and check-out, the CST must work with an Operations representative (e.g., operator or foreman) to get the new valve working properly. Figure 4-4 illustrates this process.

Another example that employs the chain of command concept is the MOC process. Before the actual change can take place (e.g., replace a valve with a different type—not "replacement in kind," which does not normally require the MOC process), the paperwork to make the change must be routed to the proper people for signatures. These are people in positions of authority who sign off on the process change.

Another example of the use of chain of command is when an FDA inspection occurs. On the day of inspection, the FDA investigator attempts to locate the top management official at the plant site. The investigator shows his or her credentials to that official and presents an FDA Notice of Inspection. The FDA investigator then conducts the inspection, accompanied by one or more plant employees (one could be a CST

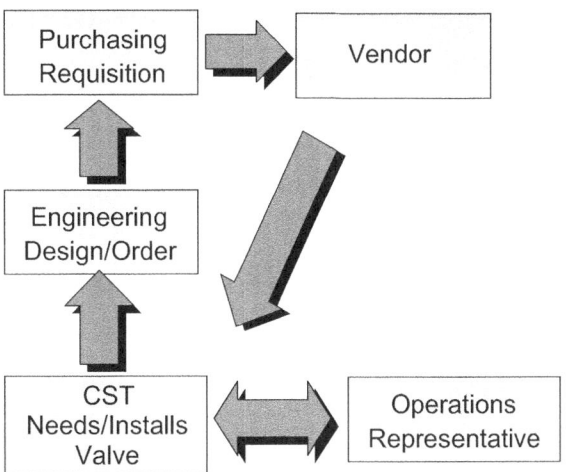

Figure 4-4. A CST coordinates ordering and installing a new valve.

acting as a liaison). After the inspection has been performed, the FDA sends members of plant top management an Inspection Observations form notifying them of objectionable conditions, if any, relating to products or processes or other violations of the Federal Food, Drug, and Cosmetic Act and related acts.

Summary

As a CST, during start-up you will interact with almost all members of the start-up team. By understanding the basic duties of the other members, you can understand their influence and knowledge in efficiently and safely starting up the plant and learn how to use their expertise to help you get your job done. By understanding the chain of command for dealing with these other groups and departments, and by understanding the ways their jobs affect yours, you can efficiently solve your own problems and help them with their jobs.

A start-up is a cooperative effort among all personnel. If everyone understands each role in the start-up, the job can be completed with fewer complications. The keys to a successful start-up are the ability to work with different people and the knowledge obtained from the initial project kickoff meeting, organizational charts, meetings, and your experience and training as a CST.

Review

4.1 A CST becomes aware that several valves from a manufacturer are not working properly. What should the CST do to prepare for the next project review meeting?

4.2 Why is it important for the CST to communicate work status to the start-up team at the project review meetings?

4.3 Which crafts would normally be included in the Maintenance department?

4.4 Why does the CST interact with so many departments, groups, and individuals during a start-up?

4.5 Why is it important for the CST to attend project start-up meetings?

4.6 Why is it important to ask for help?

4.7 What is a *contractor liaison*?

4.8 Which government agencies does the Environmental department come in contact with?

4.9 Why is the Purchasing department an integral part of a start-up?

4.10 Which discipline is the CST most likely to work with for basic process control system (BPCS)/distributed control system (DCS) work?

4.11 What type of engineer is normally involved in mass and energy balances?

4.12 Why might the Human Resources (HR) department become involved in a start-up?

4.13 What types of contractors might be involved in a start-up?

4.14 What types of vendors would be involved in a start-up?

4.15 What is meant by the *chain of command*? Why is it important?

4.16 Who can a CST ask for help?

4.17 What is the difference between *project engineers* and *process engineers*?

Recommended Reading

Sherman, R. E., and L. Rhodes, eds. *Analytical Instrumentation*. Research Triangle Park, NC: ISA (International Society of Automation), 1996.

5
Verifying and Managing Changes

The MOC Process
Maintenance, Upkeep, and Control of Project Documents

There are many reasons for making changes during commissioning and start-up. A new plant may not run as expected or designed. People involved with the project may notice that something is not correct in the documentation or with the installation of equipment. Other reasons include instrument failure, the need for additional equipment, incorrect valve sizing, the wrong construction material for a given application, and the need for range changes. Ranges may be incorrect if people think the plant can run at a higher output (production level) than expected or because operating parameters have been modified from the original design. Other changes may have occurred in the field after documents have been submitted for construction, so authorized people update documents to reflect the *as built*[1] nature of the facility.

When changes are introduced to correct these discrepancies, they must be approved and, if approved, documented. The Management of Change (MOC) process is an OSHA-mandated procedure for plants with highly hazardous materials, but many plants use MOC procedures even if they do not have these hazardous chemicals.

---

1 A plant term; refers to documentation that reflects the final project after it has been built.

Some feel this is just good practice; having a paper trail helps them make changes in an organized manner.

Operating procedures may need to be revised when there is a change in the process, whether or not the change is subject to MOC requirements. The consequences of making operating procedure changes must be fully evaluated before the change is approved, and if approval is given, the proposed change must then be communicated to all affected personnel. Additional training may be required to implement the change.

5.1 The MOC Process

The MOC process refers to a company's procedures to ensure that when a change is made to a process and/or equipment all involved parties review and discuss it, agree to make the change, complete the change, and update the relevant documentation. MOC is an extensive paper trail that helps communicate and make the changes (process, mechanical, procedural, documentation, etc.) while maintaining safe operation.

An MOC occurs when a process or equipment change is necessary. Process changes can result from changes in production rates, equipment, raw materials, experimental equipment, equipment unavailability, catalysts, and operating conditions to improve yield or quality as well as new equipment and product development. Equipment changes include changes in materials of construction, revised equipment specifications, piping rearrangements, experimental equipment, computer program revisions, and changes in alarms and interlocks.

As mentioned, the MOC process is essential for permanent changes; however, it can also be applied to temporary changes, such as when a part is not readily available and a temporary safe solution must be implemented. Employers must establish ways to detect and manage temporary changes, not just permanent ones, and they must determine and monitor time limits for temporary changes. MOC procedures should ensure that any equipment used on a temporary basis is returned to its original or designed condition after the situation is resolved.

Employers should have a form or clearance sheet to facilitate the processing of changes through the MOC process. Changes in process flow diagrams (PFDs), piping and instrumentation drawings (P&IDs), electrical drawings, loop diagrams, cause-and-effect matrices, alarm and control point settings, and standard operating procedure (SOP) documents must be noted so these revisions can be made permanent when the drawings and procedure manuals are updated. See Chapter 3 for examples of these types of documents. Even a simple change, such as replacing a pump with a flow control valve, impacts many documents, such as P&IDs, PFDs, loop diagrams, training materials, SOPs, electrical drawings, and possibly stores inventory (see Figure 5-1).

There is no standard procedure for an MOC process, although software can be purchased, or consultants hired to set up an MOC program for the plant. Depending on the facility, MOC forms are required for a proposed change. These forms, when filled out by the appropriate people, provide information that describes the change, why the change is necessary, how much it is going to cost, and the impact on the project (e.g., is downtime necessary, will rates be increased, or is this rectifying an unsafe condition?). The forms also include placeholders for different types of personnel, by department, within the organization to sign off on the change.

If a process or mechanical change is required, a design change (re-engineering) usually must be completed before the change can occur. In such a case, specification sheets must be changed and bids obtained for materials to be purchased. At many plants, a "replacement in kind" does not constitute a change that requires an MOC form to be filled out, so if a pump fails, for example, it can be replaced quickly without paperwork delay.

During or not too long after the change is completed, other paperwork (described in Chapter 3) must be updated (redlined) to reflect this and other changes. If a plant

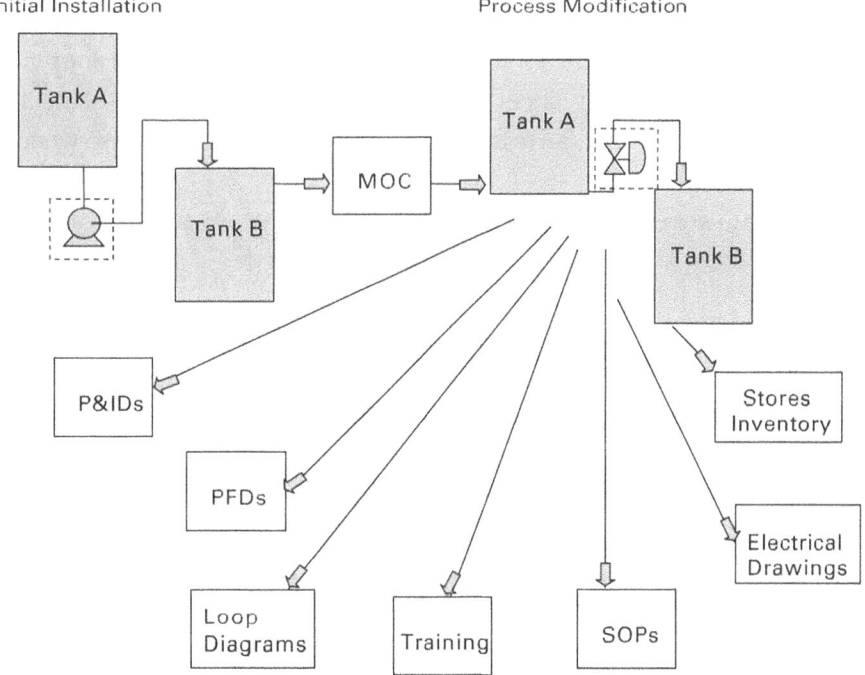

Figure 5-1. MOC: documents affected (pump to control valve change).

does not have a system to update all relevant paperwork, many drawings may not reflect the change in the plant and will thus lack "as-built" status. As a start-up team member, you can help ensure that redlines are "picked up"[2] by the Drafting department or other entity responsible for doing this.

Any documents the CST works with that must be corrected because of errors, process changes, operating changes, or programming changes should be redlined and reflect the as-built status. Even if the document is not related to the instrument and electrical (I&E) scope, if the CST sees an error, he or she should bring it to the attention of the project manager (PM) or other designated authority (possibly someone in Document Control). The project will have MOC standards by which this should occur (i.e., the chain of command and the procedures for making the change). The CST and the rest of the start-up team should be aware of how to mark up documents.

Referring back to Table 3-1, the documents a CST usually encounters that require changes to be redlined are indicated by the CST Redlines column. If the CST finds any errors, regardless of whether they are instrument-related, he or she should use the MOC process to initiate the change if it is determined that the MOC process should be followed.

A P&ID or loop sheet may need to be redlined, for example, if a control loop is shown improperly or there is a change in how the loop is to function (e.g., change from two separate proportional-integral-derivative (PID) controllers to PID/cascade control). The plant may have its own standard for showing control loops on P&IDs or use the ISA standard. A loop sheet may need to be redlined to indicate an error in input/output (I/O) assignment (termination on the process automation system).

Calibration data sheets may need to be redlined if, for example, a range change is made when a different size pump ends up being installed.

Electrical wiring diagrams may need to be redlined if, for example, a loop is moved to a different circuit or an inspection of a wire termination identifies an error on the original drawing.

In some cases, such as a start-up, MOC may be less formal, and, in some situations, parts of these procedures may be relaxed depending on the nature of the materials being manufactured (nonhazardous) and the need to make changes quickly.

---

2 A plant term that means caught and corrected.

GRACE Davison
LAKE CHARLES PLANT

MANAGEMENT OF CHANGE CHECKLIST

MOC-1   PAGE 2 OF 2

MOC NUMBER: XP9907

PART THREE: THE AREA MOC COORDINATOR IS TO DETERMINE WHICH OF THE FOLLOWING REVIEWS / REVISIONS / ACTIONS ARE NECESSARY AND TO DESIGNATE THE RESPONSIBLE PARTY. RESPONSIBLE PARTIES ARE TO CONDUCT THE APPROPRIATE REVIEW(S), MAKE APPROPRIATE REVISIONS, OR TO CONDUCT THE APPROPRIATE TRAINING.

SAFETY, HEALTH, ENVIRONMENTAL & DESIGN REVIEWS	REVIEW REQUIRED YES / NO	RESPONSIBLE PARTY	DATE REVIEW COMPLETE	REVIEWER
PROCESS SAFETY REVIEW:	☐ ☐	JULIE SMITH		CHANGE BEING REVIEWED
OCCUPATIONAL SAFETY / INDUSTRIAL HYGIENE:	☐ ☐	JULIE SMITH		CHANGE BEING REVIEWED
ENVIRONMENTAL:	☐ ☐	GARY WHITE		CHANGE BEING REVIEWED
TECHNICAL DESIGN:	☐ ☐	PHIL BLACK		CHANGE BEING REVIEWED

COMPLETE = ACTION ITEMS WITH IMMEDIATE IMPACT ARE RESOLVED AND PLAN IS IN PLACE TO ADDRESS LONG TERM ITEMS

PROCEDURES / INSTRUCTIONS REVISED TRAINING	REVIEW REQUIRED YES / NO	RESPONSIBLE PARTY	DATE REVIEW COMPLETE	REVIEWER
OPERATING PROCEDURES AND INSTRUCTIONS:	☒ ☐	SELECT ONE		CHANGE BEING REVIEWED
MAINTENANCE PROCEDURES AND INSTRUCTIONS:	☐ ☐	GARY BERNARD		CHANGE BEING REVIEWED
LABORATORY PROCEDURES AND INSTRUCTIONS:	☐ ☐	DALE JONES		CHANGE BEING REVIEWED
CONTRACTOR INSTRUCTIONS:	☐ ☐	LOUIS AARON		CHANGE BEING REVIEWED
EMERGENCY RESPONSE:	☐ ☐	GARY SMITH		CHANGE BEING REVIEWED
OTHER:	☐ ☐			

PROCESS SAFETY INFORMATION REVISED	REVIEW REQUIRED YES / NO	RESPONSIBLE PARTY	DATE REVIEW COMPLETE	REVIEWER
TECHNICAL DRAWINGS (P&IDs, PFDs, Loop Sheets, Elect.):	☐ ☐	PHIL WHITE		CHANGE BEING REVIEWED
DCS DOCUMENTATION AND PROGRAMING	☐ ☐	CODY COOK		CHANGE BEING REVIEWED
RELIEF SYSTEM DOCUMENTATION:	☐ ☐	MICHAEL WAITS		CHANGE BEING REVIEWED
MSDS:	☐ ☐	KEN JOHNS		CHANGE BEING REVIEWED
PROCESS LIMITS:	☐ ☐	KEN JOHNS		CHANGE BEING REVIEWED
MECHANICAL DESIGN LIMITS:	☐ ☐	MICHAEL WAITS	6/2/99	CHANGE BEING REVIEWED

COMPLETE = REVISIONS SUBMITTED FOR DOCUMENT UPDATE

	PSSR REQUIRED YES / NO	RESPONSIBLE PARTY
PRE-STARTUP SAFETY REVIEW	☐ ☐	Jose M. Brown

PART FOUR: CHANGE AUTHORIZATION   THIS CHANGE HAS MET THE APPROPRIATE REVIEW REQUIREMENTS AND HAS BEEN APPROVED

AREA MOC COORDINATOR: HOLD FOR AUTHORIZATION   DATE:

REV: 02/01/1999...dmc

Figure 5-2. MOC form.
Source: Courtesy of Grace Davison, Lake Charles, LA.

At one plant, MOC for start-up is not the same as it is for normal day-to-day changes. If the contractor or plant engineer discovers a problem, plant personnel may do the engineering to solve it and then submit the change. If the change is small and can be accomplished by the contractor, then the plant engineer can approve it. An example is a range change for an instrument. If the plant engineer thinks the PM's approval is required for the proposed change, particularly if it is a high-cost item or a major process change requiring a hazard and operability study or process engineering, then it has to go the MOC route.

MOC takes many forms. When it is done well, it creates a paper trail. This process may seem like a lot of red tape, but in hazardous industries it helps keep safety from being compromised. Several vendors offer electronic means for documenting MOC, but if a change affects the documents described in this chapter, then a "canned" or "out-of-the-box" solution is probably not feasible. It takes thorough integration of different kinds of software applications to document MOC electronically, and the software used to create plant documents may not be compatible with a vendor's application. Therefore, many documentation changes must be made manually.

Figure 5-2 is an example of an MOC form used when initiating a process change. The CST working in older plants (and plants that are being upgraded or expanded) should be aware that documents, including the P&ID, are often out of date and incorrect. Therefore, you should be aware of possible problems, make note of errors, and report incorrect documentation, which will help get the documents updated.

Proper MOC in industrial facilities and processes is critical to safety. OSHA regulations govern how changes are to be made and documented for facilities with highly hazardous chemicals. The main requirement is that a thorough review of a proposed change be performed by a multidisciplinary team to ensure that as many viewpoints as possible are used to minimize the chances of overlooking a hazard. As discussed in Chapter 2, MOC is one of the components of process safety management.

5.2 Maintenance, Upkeep, and Control of Project Documents

In many plants, the Engineering department maintains the originals of documents and drawings. This department often has a group working with or for them, designated as *Document Control*. During a project, Document Control personnel are responsible for maintaining up-to-date documents by ensuring that proper procedures are used to mark up (redline) changes on originals and that these documents are updated and added to the electronic document management system (EDMS) with the proper revision information and approvals (signatures). Other plant departments are responsible for maintaining their own copies, which may not be up to date. This must be checked periodically (with Document Control's version), especially if one has made a copy (paper or electronic).

For handy reference, loop sheets and electrical wiring diagrams are typically found in I/O rooms and in other cabinets and enclosures. Again, this can be a problem because you do not know if you are looking at the latest version. Because the practice of maintaining documents and drawings is dictated by plant policy, it leads naturally to the topic of MOC.

Many plants and engineering firms have adopted the common practice of using different colors of pencils or pens to manually mark up documents and drawings. Green is used for deletions or changes, red is used for additions, and blue is used for comments to the CAD operator. This allows the CAD operator to easily make the changes in the program they use to produce the documents. You should follow these practices and know how to make changes to drawings using standard symbols so the CAD operator will understand what changes to make. Once the changes are made and approved, the Engineering department should get the new copies distributed to the appropriate personnel in a timely manner.

To avoid confusion, older and out-of-date versions of drawings and documents should be discarded properly. These are often shredded because of the confidential nature of process information contained within. As described in Chapter 3, clouds on drawings represent changes from the previous document revision, so you can compare one set of documents against another to determine which is the latest and discard the old one safely. Chapter 3 provides a discussion and example of how to understand the information provided in a P&ID title block (see Figure 3-1).

The CST working in older plants (and plants that are now being upgraded or expanded) should be aware that documents, including the P&ID, are often out of date and incorrect. Therefore, you should be aware of the possible problems with having outdated documents, making note of discrepancies and reporting your findings to help get the documents updated accordingly.

Summary

Procedures for change management associated with processes vary from plant to plant and company to company. Differences include document types and document locations. The CST should be aware that there are differences depending on the type of industry and plant he or she works in.

This chapter discussed why changes occur and how the CST must work within the MOC process to get the job done safely during a project's commissioning and start-up phases.

Knowing how to mark up drawings will ensure that the latest changes are documented. This will prevent any confusion that might occur when what is actually in the field does not match the drawings.

Review

5.1 Why is it important to follow good Management of Change (MOC) procedures during start-up activities?

5.2 Describe which documents would need to be updated if a new pipeline with a valve and a pump were added to a process.

5.3 Describe a change that involves adding a new chemical that would initiate the MOC process.

5.4 What is the red pencil used for when marking up a piping and instrumentation drawing (P&ID)?

5.5 Why is it important for the CST to understand MOC procedures?

5.6 Who is responsible for generating the spec sheets?

5.7 What type of information is on an ISA standard spec sheet?

5.8 What is *redlining*?

5.9 What are *as-builts*?

5.10 Is there an ISA standard document for loop checking?

5.11 What does the Gantt chart have to do with calibration?

5.12 Why is good record keeping, including MOC, important?

Recommended Reading

Sands, Nicholas P., and Ian Verhappen, eds. *A Guide to the Automation Body of Knowledge.* 3rd ed. Research Triangle Park, NC: ISA (International Society of Automation), 2018.

6
Personnel Requirements and Responsibilities

Staffing and Overtime
Training
Emergency Contacts

As a CST, you have probably discussed the start-up plan in general terms during meetings with the start-up team (see Chapter 4). Before and during start-up, you and your associates (other instrument and electrical—I&E technicians) must deal with many related personnel, staffing, and training issues, as well as issues related to specific assignments and responsibilities. These issues are common to, and important for, all the departments, groups, and disciplines involved in the start-up.

As always, it is important to understand that all plants are different and that personnel decisions, including the CST's workload, training, assignments, and responsibilities, will be dictated by plant management and reflect the plant's operating philosophy.

6.1 Staffing and Overtime

The number of I&E personnel needed for a start-up depends on such things as plant size, plant coverage required (and any limits on hours worked), the system start-up order (discussed in Chapter 8), and the timetable for the start-up.

Although long hours (which should be expected) may place a strain on one's family life, as a CST you may find start-up worthwhile not only because it is an exciting learning experience but also because of the monetary benefits of working overtime.

Depending on the company, the location of the plant (United States or overseas), and the union contract, overtime practices vary. The US Department of Labor Fair Labor Standards Act states in the "Basic Wage Standards" section: "Nonexempt workers must be paid overtime pay at a rate of not less than one and one-half times their regular rates of pay after 40 hours of work in a workweek."

Overtime laws vary by US state. In cases where an employee is subject to both the state and federal overtime laws, the employee is entitled to overtime according to the higher standard (i.e., the standard that will provide the higher overtime pay). Extra pay for working weekends or nights is a matter of company policy or, in the case of a union shop, the agreement between the employer and the employees' representative.

The US Department of Labor Contract Work Hours and Safety Standards Act (CWHSSA) requires contractors and subcontractors on most federal contracts over $100,000 for services or construction to pay laborers and mechanics at least one and one-half times their basic rate of pay for all hours worked over 40 in a workweek. CWHSSA also applies to most federally assisted construction contracts.

Relevant Acts

US Department of Labor, Wage and Hour Division
- WH Publication 1318, *The Fair Labor Standards Act of 1938, as Amended*, US Department of Labor, Wage and Hour Division, revised May 2011
- WH Publication 1432, *Contract Work Hours and Safety Standards Act, as Amended*, US Department of Labor, Employment Standards Administration, Wage and Hour Division, revised April 2009

6.1.1 Division of Responsibility

If the plant is large and many loops must be installed and checked, many I&E personnel may be required. They usually work in pairs throughout the facility. Conversely, for a small installation, or when only part of the plant is being started up, only a few I&E personnel may be required. In this case, I&E personnel's responsibilities may be broadened to include other activities such as calibration, configuration, and troubleshooting.

When a fairly large group of I&E personnel are involved in a start-up, the CST may be asked to be the designated lead technician (possibly titled *technical lead*). The technical lead is responsible for attending meetings to report on plant and start-up status, assigning and following up on work the I&E team is performing, and communicating information to the I&E team. This is in addition to the technical lead's normal job of installation, loop checking, configuration, and troubleshooting.

6.1.2 Task Assignments and Responsibilities

A supervisor will notify the CST and his or her coworkers of their respective assignments and responsibilities for the impending start-up. These duties can be correlated with the project Gantt chart and with all meetings that have occurred involving the start-up team.

As a CST, your start-up duties are probably not carved in stone because start-up is a fluid process. When equipment has never been run before or a new process comes online, unexpected things happen. Jobs and responsibilities may change from day to day, or even hour to hour, which means you must be patient, adaptable, and cooperative. You may have someone working with or for you, which means you must communicate clearly with him or her as jobs are assigned, completed, or changed, so workflow is smooth and projects are accomplished in a safe, efficient manner. You may be a mentor for a coworker or subordinate, so your actions should reflect a safe and consistent policy that the apprentice can learn and follow.

With assignments comes responsibility. A CST should identify, in detail, what the job assignment is; who your contacts are; who to go to for help; how the job impacts the start-up, plant operations, and people; and when the job must be completed. You may want to make a checklist to coordinate these activities (see Figure 6-1).

```
Name_____        Date_____

Task Checklist: (check off when complete)

1. Check Receiving department for replacement valve for steam preheater.

2. Complete calibration of pH meter for tank 706.

3. Return extra cable to electrical contractor's trailer.

4. Call Performance Technology Inc. for help with dP cell.

5. Complete wire tags in TCB-104.

6. Pick up debris in Area 300.
```

Figure 6-1. Example task checklist.

If you are fully responsible for a certain job, you must communicate to the appropriate personnel when the job is complete. You are responsible for all aspects of the job, including its quality, so be sure that the job is finished and done well. The equipment, instruments, and software must work properly when you hand them over for operation. Also, be sure that all materials used during the job have been properly cleaned up.

6.1.3 Scheduling

When the plant first starts up, personnel may be asked to monitor plant activities around the clock. If 8-hour shifts are the rule, personnel may be asked to volunteer for a certain shift—day, evening, or night. Other plants work 12-hour shifts at a time, day or night. Whatever the schedule, if people are not asked to volunteer for a certain shift, they will simply be assigned a shift. If there is more than one CST per shift, management may determine the assignments.

Figure 6-2 shows a typical shift schedule. Each shift usually has a group consisting of several people from the Operations, Maintenance, and Engineering departments.

If the plant has a union contract and the CST is a member of the union, overtime will be paid according to the contract. Personnel will also be offered overtime based on the union rules (e.g., a more senior union employee may be asked to work overtime before a junior member). For nonunion shops, other provisions may be made to

12 Hour Shift Schedule - 2010

Shift		M	T	W	T	F	S	S	M	T	W	T	F	S	S	M	T	W	T	F	S	S	M	T	W	T	F	S	S
James Smith George West	A	D	D	D	D	–	–	–	–	–	–	–	N	N	N	N	–	–	–	D	D	D	–	N	N	N	–	–	–
Tom Alexander Otis Granger	B	N	–	–	–	D	D	D	–	N	N	N	–	–	–	D	D	D	D	–	–	–	–	–	–	–	N	N	N
Logan Elrod Etzio James	C	–	–	–	–	N	N	N	N	–	–	–	D	D	D	–	N	N	N	–	–	–	D	D	D	D	–	–	–
Paul Allen Will Jones	D	–	N	N	N	–	–	–	D	D	D	D	–	–	–	–	–	–	–	N	N	N	N	–	–	–	D	D	D
D = Days													Jan 1	2	3	4 B1	5 B2	6 B5	7	8	9	10	11 B3	12 B4	13 B5	14	15	16	17
N = Nights		18 B1	19 B2	20 B3	21 B4	22	23	24	25 B1	26 B2	27 B5	28	29	30	31	Feb 1 B3	2 B4	3 B5	4	5	6	7	8 B1	9 B2	10 B3	11 B4	12	13	14
– = Off		15 B1	16 B2	17 B5	18	19	20	21	22 B3	23 B4	24 B5	25	26	27	28	Mar 1 B1	2 B2	3 B3	4 B4	5	6	7	8 B1	9 B2	10 B5	11	12	13	14
		15 B3	16 B4	17 B5	18	19	20	21	22 B1	23 B2	24 B3	25 B4	26	27	28	29 B1	30 B2	31 B5	Apr 1	2	3	4	5 B3	6 B4	7 B5	8	9	10	11
		12 B1	13 B2	14 B3	15 B4	16	17	18	19 B1	20 B2	21 B5	22	23	24	25	26 B3	27 B4	28 B5	29	30	May 1	2	3 B1	4 B2	5 B3	6 B4	7	8	9
		10 B1	11 B2	12 B5	13	14	15	16	17 B3	18 B4	19 B5	20	21	22	23	24 B1	25 B2	26 B3	27 B4	28	29	30	31 B1	Jun 1 B2	2 B5	3	4	5	6
		7 B3	8 B4	9 B5	10	11	12	13	14 B1	15 B2	16 B3	17 B4	18	19	20	21 B1	22 B2	23 B5	24	25	26	27	28 B3	29 B4	30 B5	Jul 1	2	3	4
		5 B1	6 B2	7 B3	8 B4	9	10	11	12 B1	13 B2	14 B5	15	16	17	18	19 B3	20 B4	21 B5	22	23	24	25	26 B1	27 B2	28 B3	29 B4	30	31	Aug 1
		2 B1	3 B2	4 B5	5	6	7	8	9 B3	10 B4	11 B5	12	13	14	15	16 B1	17 B2	18 B3	19 B4	20	21	22	23 B1	24 B2	25 B5	26	27	28	29
		30 B3	31 B4	Sep 1 B5	2	3	4	5	6 B1	7 B2	8 B3	9 B4	10	11	12	13 B1	14 B2	15 B5	16	17	18	19	20 B3	21 B4	22 B5	23	24	25	26
		27 B1	28 B2	29 B3	30 B4	Oct 1	2	3	4 B1	5 B2	6 B5	7	8	9	10	11 B3	12 B4	13 B5	14	15	16	17	18 B1	19 B2	20 B3	21 B4	22	23	24
		25 B1	26 B2	27 B5	28	29	30	31	Nov 1 B3	2 B4	3 B5	4	5	6	7	8 B1	9 B2	10 B3	11 B4	12	13	14	15 B1	16 B2	17 B5	18	19	20	21
		22 B3	23 B4	24 B5	25	26	27	28	29 B1	30 B2	Dec 1 B3	2 B4	3	4	5	6 B1	7 B2	8 B5	9	10	11	12	13 B3	14 B4	15 B5	16	17	18	19
		20 B1	21 B2	22 B3	23 B4	24	25	26	27 B1	28 B2	29 B5	30	31																

Figure 6-2. Typical shift schedule.

compensate personnel. For example, an employee might be offered compensatory (or "comp") time, which means he or she can take paid time off at another time to make up for extra hours worked.

As mentioned, some start-ups may go on for a few weeks, whereas others last as long as 1 year and beyond. It is important to understand this from the beginning because people may get tired and frustrated. Many companies use enterprise resource planning (ERP) software to plan, schedule, and manage work. (Figure 4-3 illustrates the relationship between the enterprise network [Levels 3 and 4] and the process automation system [Level 2].) Such comprehensive systems do not permit deviations from the plan. Therefore, to stay on schedule, members of the start-up team are often expected to make sacrifices and work long hours. However, in the long run, the experience and the monetary benefits gained from overtime, if offered, can make start-up a rewarding experience. Successfully working as a team to get a new plant online is also a great accomplishment.

As a CST, the quantity of work required of you will vary throughout the start-up, but usually you will have to be at the plant even during idle times, for example, when a system or part of the plant is not ready to start up. Reasons for a delay resulting in idle time include the following:

- Errors in piping or electrical runs
- Incorrect piping materials
- Delays in the receipt of materials or equipment required for installation
- Failure of system(s) during a factory acceptance test (FAT), site integration test (SIT), or site acceptance test (SAT)
- Changes in project scope, that is, unscheduled changes to the project (e.g., the realization that more production capacity is needed in an area of the plant, which results in changes in design and equipment)
- Safety incidents (accidents)
- Environmental issues (failure of systems to perform or unplanned release)

After the plant is started up and running well, plant management may still want to staff the plant with experts, such as yourself, around the clock in the event of problems.

Even when the plant is running smoothly, the CST may be involved with availability and performance guarantee tests, which may be a final requirement before releasing the vendor or paying the final bill. See Chapter 3 for more on this topic. Acceptance test data and documentation must be prepared and analyzed to ensure, for example, that equipment performs as designed or that the plant produces the product to the design specifications. Therefore, even after the plant has been successfully started up, the CST and others may be asked to continue to assist with this type of work. This help may be in the form of working different shifts (e.g., night shift instead of day shift), extra hours, and/or weekends.

6.1.4 Workload and Priorities

At times during the start-up there may not be much to do, and it may even get boring, so you should look for additional work to do. For example, as a CST you might talk with the board operator to find out if anything is not performing as expected, such as a loop that is not controlling well. Or you could look at plant documentation, as discussed in Chapter 3, to understand the plant and the process more fully.

Most of the time, however, there will be plenty for you to do during the start-up. Therefore, your workload must be prioritized. Your priorities may be set by the project manager (PM), operations manager, or area foreman, but you must also discuss them with the lead CST and other CSTs.

The project Gantt chart will indicate general priorities, but each of these tasks includes many subtasks. For example, the Gantt chart may show a line item named *BMS Online*, but before a burner management system (BMS) is started up, it is likely that the instruments associated with the BMS must be calibrated and then loop checked. If, in this example, gas or oil pressures differ from the design, the PM will be prompted to request that a factory representative come to the plant to do additional calibrations, checks, or inspections. Typically, a factory representative makes this type of visit if the plant is a new installation or if the vendor offers this service as part of the purchase.

Before the representative arrives, and during his or her visit, the CST might need to make sure that the installations of the electrical system, pneumatic system, and process automation system (PAS) have been completed correctly. Personnel may need to be trained to understand how the BMS should operate. Loop checks also must be completed.

Problems with installation, training, and loop checking may delay starting up the unit. Other work that should be occurring at this time may be neglected as personnel concentrate on the immediate problem. This could affect the project schedule and impact the cost of the project.

All these issues must be communicated to representatives of management, who will probably be working daylight hours when the lead CST is on the night shift. Problems must be documented accurately; if they are serious enough, the lead CST might have to call (and wake up) the managers, tell them what is going on, and possibly ask them to make a decision. More personnel or money or a shift in workload and priorities may be required to solve the problem. The PM must get involved to keep the project on schedule and within budget.

6.1.5 Coverage during Different Start-Up Phases

Fewer people may be necessary during commissioning and during the early phases of the start-up because equipment might not be installed yet or wiring might not be complete. During this period, commissioning personnel might only work 8-hour daylight shifts. As the time approaches for the start-up to begin—and as production deadlines near—managers often decide to accelerate completion by having contractors and plant personnel work overtime and/or by increasing the size of the workforce. As most plants do not let people work more than two shifts or 16 hours at a time because of union contracts or safety concerns, the schedules of qualified CSTs must be organized to cover all plant start-up needs. If there are only a few qualified CSTs, then the lead CST may be asked to work more than 8 hours per day and even weekends or nights.

6.2 Training

A start-up requires that people work around the clock, but there are physical as well as legal limitations to the number of hours a person can work. Therefore, it is important to have more than one person available to do a task well. This usually means having people on hand who can perform more than one set of tasks.

Every company should have a cross-training program in place so people are ready to work in case of heavy start-up demands, absenteeism, or understaffing. Training may take place off-site at a school (e.g., vendor location) or on-site using qualified instructors, or self-training methods (videotape and audiotapes or computer-based training) may be offered. On-the-job training may also be offered, for example, by a more experienced CST or a control systems engineer (CSE).

As mentioned earlier, a CST should be well versed in many plant activities, including calibration, wiring and installation, loop checking, configuration, and troubleshooting. You may have more experience in one of these activities than another. If you feel you are lacking experience or skill in any of these areas, you should bring this to your supervisor's attention. If warranted, he or she can then make provisions for additional training.

However training is accomplished, more than one person should have the skills to perform a task during a start-up, especially if the task must be "manned" around the clock. It is up to plant management to ensure that a particular task gets covered—not only in case more than one person is needed, but also as a security measure for the plant in the event of absence or personnel turnover.

As a CST, you may receive cross-training in the installation, configuration, troubleshooting, and maintenance of the PAS and its subsystems and interfaces (e.g., distributed control system—DCS, programmable logic controller—PLC, and other electronic devices and computerized equipment). People with knowledge of this field are in great demand, so it is in your best interest to take on this added responsibility.

6.2.1 Installation Training

Installation involves knowing what components have been ordered for the project and helping to assemble them and power them up. For example, before power-up, "bump tests" must be performed to ensure proper motor (e.g., pump) rotation. Baseline vibration data should be collected on all rotating equipment the first time it is operated. The CST is involved with all these tasks. Training, studying documentation, and working with suppliers will help the CST to do an effective job during this crucial phase prior to start-up. Calibration normally follows installation, and the CST should be trained for this and use the plant-approved methods to complete the job accurately.

6.2.2 Configuration Training

As discussed previously, many more devices require programming than simply the PAS. These include the following:

- PAS subsystems (DCSs/basic process control systems—BPCSs, safety instrumented systems—SISs, PLCs, third-party packaged systems—TPPSs)

- Instrumentation

- Valves and other field devices with electronic communication capability

- Interfaces (e.g., wireless and fiber optic converters, Open Platform Communications—OPC,[1] and fieldbus)

Configuration of the PAS normally requires using loop sheets and other documentation to create tags that identify wiring termination locations and to make

---

1 Note that OPC originally stood for Object Linking and Embedding (OLE) for process control. The term now stands for Open Platform Communications and refers to industry standards focused on interoperability.

connections for signal inputs and outputs. More complex configuration involves connecting inputs and outputs into control schemes that link controllers and automate the process through logic and automation sequences.

A complex loop normally refers to process control that is more complicated than single-loop proportional-integral-derivative (PID) control such as ratio, cascade, feedforward, high- or low-override, or multivariable control. An example of PID versus cascade control is shown in Figure 6-3.

The CST may be required to perform this type of work or, at minimum, assist with it and understand its purpose. Configuration may also involve creating custom graphics and configuring trend displays and alarms for the operator who must effectively run the plant. On-the-job configuration training of the PAS may be offered by a CSE or contract system integrator. This is beneficial because the training is then plant specific. A vendor course, commonly offered by the main automation contractor, often covers more material but does not include individualized training.

It is also important for a CST to know how to use tools such as the handheld communicator and software used with intelligent (smart) transmitters and final control elements and meters to check instrument signals terminated at the PAS termination assemblies. Often referred to as *field device management* (FDM) or *asset management system* (AMS) software, this software can be used to remotely configure and maintain smart instrumentation and field devices by automatically detecting these devices and adding them to the control system database. The software can compare field and

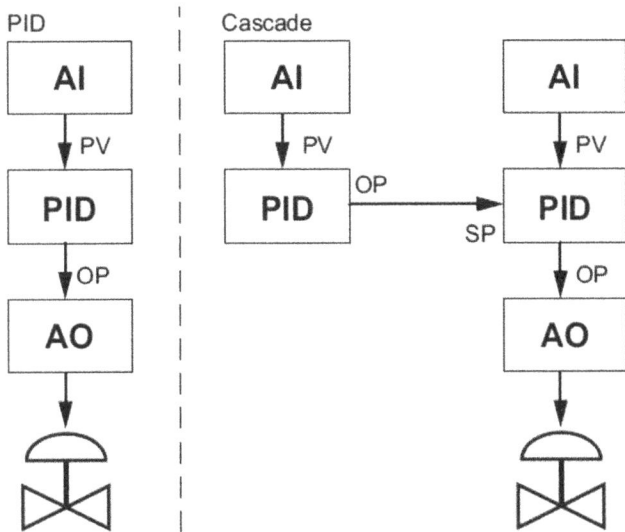

Figure 6-3. Simple (PID) versus complex (cascade) control loop example.

control system configuration; provide alarming when values or parameters do not agree, helping improve start-up activities by allowing instrumentation configuration to be downloaded and/or uploaded from the control system or the field perform; and perform partial stroke testing. It can also help with inventory management and job control, and produce certificates to ISO 9000:2015 standards.

Some plants may also be interested in using procedures and shops that use ISA-20-1981 so they can manage OSHA 1910 and ISO certification documentation for their instruments. As a CST, you may be responsible for making configuration changes at the field elements or the PAS to make sure readings are correct all the way through the loop.

Finally, training on other systems connected to the PAS will help you work on them and understand where data is coming from or going to and whether the information is accurate. For example, a TPPS may provide data to the PAS, whereas a higher level system (e.g., AMS) may get data from the PAS. If an operator at the PAS human-machine interface questions this data or an engineer on the enterprise (business) network questions data provided from the PAS, you may be asked to investigate the problem.

6.2.3 Maintenance Training

Maintenance training may include the following:

- Identify hardware components (including input/output—I/O assemblies, controllers, networks, and servers).

- Understand system diagnostics and identify failures using the system displays, system software and hardware indicators, status displays, and error codes.

- Repair system failures.

- Understand software configuration (basic I/O, soft tags, complex loops, and automation sequences).

- Troubleshoot OPC and other communication protocol problems.

- Perform system backup and restore.

- Build and commission loops (PAS software and I/O termination).

- Perform routine maintenance on hardware.

- Use PAS supplier documentation to obtain additional information to support these tasks.

To maintain equipment and instrumentation connected to a PAS, a CST must understand how to determine where in the PAS each loop or interface is connected. This includes finding the loop that the operator is having problems with, tracing it out to the PAS and field, and determining the cause of the issue. Problems could be caused by factors such as a configuration error, instrument out of calibration, physical disconnection, or hardware failure.

As a cautionary note, regardless of whether the loop is a critical loop, care must be taken before attempting any changes, however small, because some loops can cause a plant to shut down and/or create a hazardous situation. This may happen because the part of the loop (component) you are working on is part of an interlock (input or output) and the logic is set to perform some sort of function if the component you are working on meets or exceeds some limit (high, low, out of range, off-line, etc.). This writer cautions everyone who works on an active loop to refer to all documentation (e.g., interlock matrix and loop diagrams) before attempting any troubleshooting, disconnection, or simulation. As an aid, some plants mark critical loop components differently than noncritical loop components in the field (e.g., with tags or paint color).

Another maintenance function is removing electronic components; cleaning them, if necessary (e.g., pH probes or printed circuit boards); and replacing failed electronic components. A CST must know where the spare parts are and the appropriate procedures (e.g., using antistatic wrist straps) for replacing the components.

You must also follow other maintenance procedures, such as returning failed components and ordering parts for the Plant Stores department or from the vendor with a purchase order.

Keeping track of equipment is an important part of maintenance. As discussed earlier, an AMS in conjunction with a plant program that tracks equipment status and creates work orders has an impact on the CST. You must understand this system and work within it.

6.2.4 Other Training and Equipment

CSTs may be offered other training, such as training in using specialized equipment for calibration, both in the shop and in the field, and in newer technologies such as fieldbus and fieldbus devices, intelligent (smart) transmitters, continuous emission monitoring systems, supervisory control and data acquisition systems, wireless, OPC, and fiber optics.

Depending on the plant atmosphere, some types of tools, even common ones such as wrenches, must be nonsparking (explosion-proof). The CST must be trained to know when such tools are required.

Training to use test tools back at the shop on the bench may be required because testing may involve the use of specialized equipment such as oscilloscopes, circuit board testers, and even microscopes.

Relevant Standards

American National Standards Institute

- ANSI/ISA-61010-1-2012, *Safety Requirements for Electrical Equipment for Measurement, Control, and Laboratory Use – Part 1: General Requirements*, Second Printing 15 July 2015

- ANSI/ISA-12.12.03-2011 (2015), *Standard for Portable Electronic Products Suitable for Use in Class I and II, Division 2, Class I Zone 2 and Class III, Division 1 and 2 Hazardous (Classified) Locations*

International Society of Automation

- ISA-82.03-1988, *Safety Standard for Electrical and Electronic Test, Measuring, Controlling, and Related Equipment*

Underwriters Laboratories

- UL 1203, *Explosion-Proof and Dust-Ignition-Proof Electrical Equipment for Use in Hazardous (Classified) Locations*, October 28, 2009

- UL 60745-1, 4th Edition, *Hand-Held Motor Operated Electric Tools – Safety – Part 1: General Requirements*, November 28, 2016

There are many different types of analyzers, and each is very specialized. If you will be responsible for installing or maintaining analyzers, your training will probably occur at the vendor site or with a vendor representative at the plant site. Analyzers are often connected to specialized computers and servers or require connections different than standard copper wiring to provide the information to the PAS (e.g., OPC or Modbus). In addition, analyzers often have to be run with "standards" against which their calibration and accuracy are tested so that when they measure process materials, the readings they provide are correct.

Chapter 6 – Personnel Requirements and Responsibilities   133

Relevant Recommended Practice

International Society of Automation

- ISA-RP76.0.01-1998, *Analyzer System Inspection and Acceptance*

Although they are becoming scarce, data recorders may still be required in a new plant (because of environmental regulations) or may still be in use at an older plant site. You may be involved with this equipment's upkeep. Therefore, you must be trained in its operation and maintenance.

Finally, CSTs should be trained in computer software applications normally used on a laptop to connect to and communicate with devices (e.g., a PLC) in the field.

If any of these tasks are your responsibility and you require additional training to perform them, you should notify your immediate supervisor.

6.3 Emergency Contacts

As a CST, it is your responsibility to review the list of emergency contacts in preparation for start-up. This list should include the names and numbers of the individuals

---

Emergency Communications Guide

Contacts should be made in this order:

1. Call 4444 (plant emergency response center)
2. Activate paging system
3. Contact fence-line neighbors listed below

Name	Phone Number
ABC Chemical Company	555-1234
EF Distributors	555-4231
GHIJ Engineering	555-3352

4. Contact local industries reporting agency (CAER)
    555-2260   555-2200 FAX

5. Contact Environmental agency
    555-4000   555-4400 FAX

6. Call Managers listed below

John James, Plant Manager	555-3890
Janice Brown, EHS Manager	555-0008
Joseph Jones, Operations Manager	555-7839
Samuel Tone, Personnel Manager	555-8270

Figure 6-4. Example emergency contacts list.

or organizations to be called in the event of fire, chemical release, injury, mechanical failure, or disruption of production. Figure 6-4 is an example of an Emergency Communications Guide, set up as a standard operating procedure.

This list should be posted in a location where all personnel can see it and a copy should be given to every person on the start-up team.

Summary

The number of I&E and CST personnel needed for a start-up depends on such things as plant size, plant budget, and the timetable for start-up. Responsibilities such as calibration, installation, configuration, loop checking, and troubleshooting will be divided up among the personnel who are qualified to do this work. Factors influencing the completion of this work are staffing, overtime, varying workload (from commissioning through completion of start-up), the number of qualified people available (which can be increased through an effective cross-training program), and changing priorities. Changes in priorities may be the result of unexpected problems or changing customer (plant output) needs.

As always, record keeping is important for communication and for any MOC procedures that must be adhered to.

Having emergency contacts available becomes particularly important when the start-up begins.

Review

6.1   What are some productive things a CST can do during slow (idle) times?

6.2   Why is it important to have people cross-trained in preparation for a start-up?

6.3   Why might the number of CST personnel involved in a start-up differ from plant to plant?

6.4   What is a *lead technician*, and what are this technician's responsibilities?

6.5   What jobs might a CST be involved in?

6.6   Why does the quantity of work required of the CST vary throughout the start-up?

6.7   What things must be completed before a vendor representative arrives at the plant?

6.8   What are some skills or job functions that a CST may receive cross-training in?

6.9   What safety factors should be considered in connection with the process automation system (PAS) when preparing for start-up?

6.10 How are the CST's assignments and responsibilities determined?

Recommended Reading

Berge, Jonas. *Fieldbuses for Process Control: Engineering, Operation, and Maintenance.* Research Triangle Park, NC: ISA (International Society of Automation), 2002.

7
Prestart-up Activities

Equipment and Instrumentation Installation
Quality Assurance/Quality Control Inspection
Configuration
Mechanical Completion
Pre-Commissioning
SIMOPS
CST Tools, Test Equipment, and Technology
Working with Vendor Representatives and Specialists
Commissioning

During the project phases leading up to start-up, many activities are occurring. Equipment and instrumentation may still be undergoing installation (construction phase), instruments must be calibrated, loops must be loop and function-checked, third-party packaged systems (TPPSs) must be connected and tested, and quality and safety inspections must be performed on all systems. The process automation system (PAS) may require additional programming and changes now that loop and function testing checks are occurring and mistakes are found.

The order in which prestart-up activities will be completed is communicated at the project review meetings and reflected in the project schedule (Gantt chart). The project team must be kept informed of the order in which activities shall proceed. Monthly and even daily meetings as start-up nears keep everyone up to date on project progress. Because most of the equipment or unit operations in the process are interrelated, the order of commissioning and start-up is carefully thought out to ensure that the start-up deadline can be met.

Calibration, loop checking, and troubleshooting are all CST activities during pre-start-up and will be touched on in this chapter; however, these topics are discussed in detail in the *ISA Technician Series* books by Mike Cable, Harley Jeffery, and William Mostia listed in the "Recommended Reading" section at the end of this chapter.

Figure 7-1 illustrates a generalized outline of when the activities discussed in the next two chapters occur. The construction phase of the project should be nearing completion before pre-commissioning activities begin, but, in reality, many times construction continues while pre-commissioning and even commissioning activities occur. When this occurs, extra care must be taken to avoid accidents and unintentional

Construction
- Complete construction activities
- Complete QA/QC; issue punch lists
- Agree to transfer any punch list items to pre-commissioning

Pre-Commissioning
- Receive mechanical completion certificate
- Receive procedures for completing pre-commissioning
- Complete pre-commissioning activities
- Agree to transfer any punch list items to commissioning

Commissioning
- Complete commissioning activities, including vendor call-out
- Complete cleaning, blows, and leak testing
- Complete initial fills and lubrications
- Finalize S/D logic
- Start up non-hydrocarbon systems
- Complete C&E checks and SIS testing

Start-Up
- Conduct PSSR and transfer project to Operations
- Initial introduction of hydrocarbons
- Initial ramp-up
- Live ESD testing
- Loop tuning
- Performance running and proving (BO)
- Final acceptance audit

Operation
- Post project activities
- Sustainable SS operation
- Project close-out (contracts, documentation update, work order system, lessons learned)

Figure 7-1. Final project phases.

mistakes, for example, changes to instrumentation after they have been loop checked and removal of equipment that has been cleaned. When construction and pre-commissioning and commissioning activities overlap, simultaneous operations (SIMOPS) described in Section 7.1, is an important process that must be followed.

7.1 Equipment and Instrumentation Installation

Installation detail (typical) drawings illustrate how items relevant to the CST should be installed and connected to process lines and equipment during the construction stage. These drawings were discussed in Chapter 3. Many plants have contractors representing different crafts (instrument, electrical, or mechanical) perform this work, while some plants use their instrument and electrical (I&E) group to install the instrumentation. Instrument calibration normally follows instrument installation unless the project has paid for the instrument to be factory calibrated whereby *calibration certificates* will be provided by the vendor. Even if calibration was done at the factory, calibration is often rechecked after the instrument is installed in the field.

7.2 Quality Assurance/Quality Control Inspection

Quality inspections (discussed in Chapter 1) may be occurring in different areas of the plant while other activities are occurring in another part of the plant. This is especially true if the project is fairly large. Quality inspections generally happen before loop checking to ensure the equipment has been installed properly. A quality assurance/quality control (QA/QC) form, along with other documentation, is placed in the loop folder before the loop is checked. The loop folder and documents contained within it were discussed in Chapter 3.

Depending on the size of the project, the quality/inspection group can range from one individual to several people per craft (e.g., I&E, welding, and safety). There must be at least one inspector for each category of work. Each inspector has specific expertise and must inspect his or her respective part of the project. Items that fail to meet each area's criteria to pass inspection are placed on a "punch list" of items to be addressed by the appropriate personnel. Often, the punch list is published and brought to the project review meetings so the start-up team can see what items must be addressed before the construction is complete (i.e., the mechanical completion) and the plant can be started up. Someone with the appropriate expertise is usually designated, by name, to take care of the problem so that the inspector can verify the work was satisfactorily completed, and the item can be removed from the punch list. Items relevant to the CST could end up on the punch list and the CST must address the item(s) before the start-up can commence.

The quality inspector's typical duties include:

- Comparing the installation detail with the actual equipment installation to ensure it has been done properly. This includes checking terminations, instrument air connections, and wiring tags.[1] Wiring runs and instrument installation should follow project and industry standards.

- Examining the ergonomics of the situation, for example, determining whether an installation will be easy to work on. An example of poor ergonomics is when instrumentation becomes inaccessible because it is installed high above the plant floor, or it is wedged between other pieces of equipment, or because equipment is built around it.

- Checking that support (i.e., hangers) for instruments is adequate and that equipment structures (e.g., concrete pump bases) ensure that equipment is level and have bolts or other hardware that secure the equipment properly.

- Checking that ground wires are attached and using a method to check that the grounding is sufficient, for example, compliance with 81-2012, *IEEE Guide for Measuring Earth Resistivity, Ground Impedance, and Earth Surface Potentials of a Grounding System*.

- Checking that conduits and conduit covers are closed. Verifying that seals are poured for Class 1 and Division 1 and 2 areas.

- Checking that junction boxes and other enclosures are labeled properly, powered, grounded, closed, and cooled or blanketed, where applicable.

- Checking areas for housekeeping.

- Verifying that project documentation is being used and that it is being filled out properly.

The quality inspector may add his concerns about these items to the punch list so they are brought up at a project review meeting for a discussion and decision by the PM about whether or not they should be corrected.

Typically, a loop must pass the quality inspector's approval and a loop folder must be completed before a loop can be loop and function checked.

---

1 A wiring tag is typically a physical plastic or metal tag with a wire or a loop number engraved on it.

The quality inspector will also verify, periodically, that all items are in the loop folder and are signed, dated, and checked off properly as loop and function checking progress. Document Control personnel may also be involved with this last item.

After the QA/QC process is complete, the CST and others are ready to start loop and function checks. When the loop check is complete and successful, the CST and inspector sign off on it.

7.3 Configuration

Before and during construction, control systems and other programmable devices undergo what is generally referred to as configuration. Today almost every device (e.g., instruments and transmitters) is microprocessor-based. As a result, there are many more devices that require programming than simply the PAS. PAS configuration generally means programming done for the following: input/output (I/O), interface to third-party systems, basic and complex control schemes (including interlocks and sequences), graphics, and databases for collecting process history for trends and other uses.

After the initial configuration is complete, additional programming may be required to modify control loops either on the PAS or by using handheld communicators in the field to upload or download transmitter and digital valve information. These devices may be referred to as *smart* or *intelligent* instruments.

PAS configuration creates connections to input and output devices and control access via a human-machine interface (HMI), usually by using custom graphics that have been developed, by which the plant is operated.

Using a handheld communicator or a field device management (FDM) software application in association with all intelligent/smart transmitters and final control elements such as valves (e.g., digital control valves—DVCs) involves downloading and uploading electronic data from and to transmitters and valves so their signals correspond with those of the PAS. If there is a mismatch of range, that is, if the transmitter is configured for a range different than the PAS, then readings (field signals) shown on the operator's monitor (HMI) will be incorrect.

The CST should also be aware that critical equipment such as compressors often communicate to machinery monitoring and protection systems (MMPS) and HART instruments to AMSs which then interface to the PAS. If the people responsible for using this data feel there may be a problem with it, the CST may be asked to work with them to verify the field signals to and from these systems. Additional CST responsibilities may include maintaining the MMPS and/or AMS and/or configuring signals to and from them.

Smart instrumentation often requires specialized knowledge. This author recommends all CSTs get experience working with (or perhaps training in) such instrumentation and their associated systems.

7.4 Mechanical Completion

Mechanical completion is a milestone in a project's contract where the installation/construction phase can be declared complete because all equipment has been installed (including electrical and instrumentation) and has been pressure tested. The project can be then be handed over to another entity so commissioning activities can start.

Depending on the contract between the owner company and the engineering firm, pre-commissioning may be included as part of mechanical completion.

7.5 Pre-Commissioning

The following activities are typically completed during the project pre-commissioning phase:

- **Power and grounding** – Power is supplied to all field instrumentation, equipment, and enclosures, and grounding is connected properly to all three. Starters have fuses installed and wires terminated. Lockout/tagout (LOTO) procedures are in effect.

- **PAS** – The PAS and all subsystems and interfaces (e.g., distributed control systems—DCSs, programmable logic controllers—PLCs, and any other TPPSs or single-loop controllers) are online and operational. Grounding per manufacturers' instructions is extremely important.

- **Instruments**– Instrument air, electrical, and process connections are complete. Instruments have been calibrated.[2]

- **Analyzers** – As with other instrumentation and devices for the project, analyzers must be installed and calibrated[3] before start-up occurs. Usually, a field representative from the company that supplied the analyzer is engaged

---

[2] Depending on the project and management decisions, calibration may occur at the factory before instruments are shipped or at the plant during pre-commissioning or commissioning.
[3] Depending on the type of analyzer, standard solutions are used during calibration.

to assist with ensuring the analyzer has been installed properly, is calibrated (often with standard or reference samples), and is communicating on the necessary network(s) it is intended to be connected to. An example of a continuous process analyzer is a gas chromatograph. It is important to understand what these analyzers are used for, how they are connected, how they communicate (e.g., Modbus), and the data they are providing to control the plant. You may be involved with connecting, configuring, maintaining, and troubleshooting these systems and therefore can learn a lot from the field representative, the literature supplied by the vendor, and the documentation that should accompany the device(s). Not all plants use analyzers, so you must know whether the process you are working with uses them.

- **Gas detectors** – Similar to analyzers, gas detectors must be installed and calibrated[4] before start-up occurs. Again, a field representative from the company that supplied the detectors may be engaged to assist with start-up. You may be involved with connecting, configuring, loop checking, maintaining, and troubleshooting these devices and their associated systems and therefore can learn a lot from the field representative and from the literature supplied by the vendor. These devices may not be part of the project you are involved with, especially if the plant does not have flammable or combustible gases present, but if it does, then it is important that these devices are working before chemicals are introduced to the facility, during and after start-up. Gas detectors are often associated with alarm and evacuation systems and can set off false alarms, resulting in plant evacuations, or worse, not function when needed, resulting in an unsafe environment in which a fire or explosion could occur without warning.

- **Alarm and evacuation systems** – Gas detectors often work in conjunction with alarm and evacuation systems. Again, a field representative from the company that supplied the system may be engaged to assist with start-up, and you may be involved with connecting, calibrating, configuring, loop checking, maintaining, and troubleshooting these systems.

- **Laboratory measurements** – Some plants require the use of laboratory measurements and results before start-up can take place. They may be associated with in-line and continuous samplers with analyzers or may entail a person taking a sample manually, bringing it to the lab, and then waiting for a result from a laboratory technician before deciding whether the product is good (passes the test criteria) or not. You may be involved with the installation, calibration, start-up,

---

4   Standard gases are used to calibrate.

and upkeep of an in-line sampler. The results of laboratory analysis depend on how well the plant is running and the instrumentation used to perform measurements and process control. Therefore, as a CST, you must understand how laboratory results impact your job and the start-up at hand.

- **Process lines and vessels** – Start-up strainers are installed as necessary. Lines and vessels have been leak checked. Lines and vessels are being blown out, which is normally Operations' responsibility, and/or chemically and/or mechanically cleaned.

- **Special equipment needs** – Equipment that requires lubrication is checked to ensure that it has the proper fluid levels. Pumps that require seal water or cooling have these elements available and working properly. Refractory units (equipment containing firebrick) are being cured and started up using the manufacturer's directions. Rotating equipment and motors are checked that they turn in the proper direction. Catalyst is being charged to systems which require this.

Even though some of these tasks are other disciplines' responsibility, as a start-up team member you must understand that you should not allow equipment to run or be used that is not ready. Doing so can cause equipment damage, potentially delay the start-up, and impact cost—and you might be blamed for the error.

7.6 SIMOPS

Short for simultaneous operations, SIMOPS is an important administrative control that addresses instances where two critical but necessary operations are being carried out within close vicinity of each other, and they pose a very high safety risk.

Care must be taken to manage work and ensure segregation and that safety precautions are taken. One example of SIMOPS is a situation where one party is drying out lines with high-pressure air that are interconnected to an area under construction. Companies have procedures to ensure that SIMOPS situations are handled properly. All personnel involved with the project must be aware of these procedures to ensure they participate in the project safely for themselves and all others working every day.

An example of a SIMOPS situation during commissioning may be performing calibration or loop checks in the process area while high-pressure line blows are being performed. Another example is a situation where the CSE engineer is testing PAS logic by opening and closing valves, while Maintenance personnel are cleaning out a tank. In such situations, precautions and procedures are in place that ensure safety is maintained while both of the jobs are done *simultaneously*.

Such precautions include:

- Permits with signatures of all personnel involved
- Procedures including barricading,[5] JSA, fire watch, LOTO, and clearly identifying live lines and isolations
- Communication including meetings, announcements, marked-up documentation, and postings

The CST should also be aware of situations where SIMOPS are occurring and the procedures involved. When construction and commissioning overlap, the potential for near misses increases.

7.7 CST Tools, Test Equipment, and Technology

Instrumentation and control technicians work on a range of instruments including primary control elements, transmitters, analyzers, detectors, signal conditioners, current-to-pneumatic (I/P and P/I) converters, solenoid valves, controllers, and final control elements (e.g., valves). To do this, they use hand, power, and electronic tools and test equipment such as the following:

- A test gauge, 0–30 psig (0–207 kPa), for calibrating pneumatic field instrumentation and final control elements.
- A calibrated hand pump, for simulating pressure signals for pressure and flow (dP) field instrumentation.
- Hand tools, including screwdrivers, wrenches, slip-joint pliers (or channel locks), sheet metal snips, soldering guns, and flashlights. A flashlight is useful for looking at wiring and wire tags in enclosures that are often dark. Flashlights or lights specific for "ringing out" fiber-optic cable systems are important tools to have on hand.
- A 4–20 mA current source, self-powered and with an adjustable current output, such as a multimeter, sometimes referred to as a loop calibrator. Depending on the manufacturer and model, it can measure and source milliamperes, volts, frequency, and ohms for temperature (resistance temperature detector—RTD and thermocouples), pressure, and other signals during loop check.

---

5 Taping/roping off areas with signage.

- Tools and software used for HART, PROFIBUS, and other digital systems (e.g., fieldbus) or a HART Communicator.

- Personal computer (PC) laptops or tablets and relevant software, such as Microsoft Excel for business applications, or the PAS-specific software for viewing, changing, or backing up logic or communicating with instrumentation and valves (AMS and FDM software—Chapter 6), and communicating with TPPS.

- A tachometer for measuring shaft rotation speed.

- Cables to connect to a TPPS and systems for which you will be using a meter or laptop.

- Standard chemicals or reagents (e.g., pH, process gases, and gasoline components) to calibrate analyzers and gas detectors, for example.

Make sure your equipment is suitable for working in the area of the plant in which you will be using it. Plant policies and procedures must be followed and industry standards can be reviewed if you need additional information to ensure that your tools will not compromise your or anyone else's safety.

7.8 Working with Vendor Representatives and Specialists

If during commissioning, you, the CST, have done everything possible but a problem with a loop, for example, persists and there is no one at the plant who can help, you may need to call in a vendor representative. This might be necessary for a problem with a field device, such as a valve or a transmitter, or a component or larger part (subsystem) of the PAS. You may also need to call a vendor representative or specialist (or system integrator)—which we will refer to as "specialist" in this section—if you are not trained in certain software, instrumentation, or equipment. The specialist may be helping to commission a particular system and will be very familiar with the control software, instrumentation, or equipment—and perhaps all of them. Consequently, you can get expert training and information from the specialist about how the system works and how to troubleshoot it. Take advantage of this opportunity, ask questions, and ask to do some of the work yourself while the specialist oversees what you are doing. People learn best through hands-on work, and there may be limited opportunities to work with specialists later.

A specialist can also draw on other experts and designers if he or she needs additional help or information. Being there when the specialist makes these calls can help you to better learn the system and how to make these contacts. Take notes, and get the specialist's business card for future reference.

Finally, the specialist may have access to parts or printed information within the manufacturer's or vendor's organization. The parts might be spares that must be ordered or a replacement if something breaks during start-up. For you as the CST, and for the plant as a whole, it is important that you learn about these items and make the contacts necessary to procure them.

7.9 Commissioning

The International Society for Pharmaceutical Engineering (ISPE) defines commissioning as:

> A well-planned, documented, and managed engineering approach to the start-up and turnover of facilities, systems, and equipment to the end user/customer that results in a safe and functional environment that meets established design requirements and customer expectations.[6]

Typical activities that occur during the commissioning phase of a project are:

- **Loop checks** – Loop checks involve technicians in the field and someone working at the PAS console, also known as the *human-machine interface*. Sometimes an additional technician is present where the I/O (terminations to the PAS) are located if these are not in the field near the other technician or instrument. Using the loop folder, the person at the PAS typically verifies that the loop operates properly by checking that the commanded state of the device (e.g., open valve) agrees with the feedback state of the device (e.g., valve is open), while the person in the field checks signals to and from the control room and verifies that the device performs as commanded (valve is open). Other types of loop checks involve sending a 4–20 mA signal from the field to the HMI at five different signal levels (five-point check at 4, 8, 12, 16 and 20 mA).

- **Function checks** – Function checks involve technicians in the field and someone at the PAS HMI. Function checks can be more complicated than loop checks because in addition to checking signals to/from the field devices from/to the PAS logic, the logic to control these devices is also tested. Logic may involve coordinating transmitter alarm states with valves or pumps (e.g., interlocks) or may constitute sequential control (e.g., recipe management).

---

6 International Society for Pharmaceutical Engineering (ISPE), *IPSE Baseline Guide: Commissioning and Qualification*, 2nd ed. (North Bethesda, MD: ISPE, 2019).

The following is an example of a simple function check:

- A pump starts by logic when a level is high and then turns off when the level is low. The CST simulates low- and high-level signals while someone at the HMI witnesses that the pump starts and stops.

- **Interlocks** – Interlocks also require function checks but depending on the safety integrity level (SIL)[7] rating, interlocks may be executed within a BPCS or a plant safety system, also known as an SIS, or in both.

  - The following is an example of a more complex function check involving an SIS: A "trip" is executed when something critical triggers multiple pieces of equipment to stop and/or multiple valves to open or close to ensure that a safe situation is maintained and a dangerous situation is averted. Again, the CST and qualified personnel simulate these situations and witness that the correct actions occur. Documentation such as control narratives and cause-and-effect (C&E) diagrams are used to check this functionality (the SIF), and sign-offs are required after the testing is completed successfully. If the tests do not execute successfully because of a problem with the SIF logic, a qualified person, for example, the *certified functional safety professional* (CSFP), can fix it by doing some reprogramming. If the problem is instrumentation, then the CST may be involved.

- **Plant assistance** – Vendor/manufacturer or service representatives for certain pieces of equipment are present if they are required to be. Personnel from sister plants who have experience in the process or in commissioning and start-up may also be asked to assist. Internet searches for information are another way to obtain assistance. Check if any of the equipment or systems in the project have contracts set up whereby one can log in to the vendor website to obtain additional information electronically. Finally, phone numbers of important contacts should be available, for example an 800 number for technical assistance on special equipment and/or systems (e.g., the PAS).

- **Operations personnel** – Plant operators are typically assigned to assist the CST and I&E personnel during loop and function checks. You should stay in direct communication with the plant operator during this fast-paced and sometimes unpredictable part of the job. Permits are required when stroking valves and starting machinery. Make sure your two-way radio is charged.

---

7 A relative level of risk-reduction provided by a safety instrumented function (SIF) that is a combination of logic (the SIL), the logic solver (SIS controller), and instrumentation and field devices connected to the SIS and dedicated to this application.

- **Recordkeeping and Management of Change** – These work processes are important throughout this phase of the project. Once commissioning is complete for one area of the project, documentation such as completed and signed-off loop folders, C&E matrices, and control narratives are a record of what equipment and loops have been "sold" and are now Operations' responsibility.

Regardless of the order of loop and function checking followed, technicians and inspectors should never get hung up on a faulty loop but rather move on to other loops, returning to the faulty one when the repair is made and the loop is again ready to be checked. This ensures that while people are waiting for one loop to be fixed, they are working on something else. Waiting around wastes a lot of time and may even delay the project.

The loop check coordinator may be the project manager (PM), a control systems engineer, a project or process engineer, or a CST. This person coordinates the loop checking activities that involve the Operations, Maintenance, and Engineering departments, as well as contract personnel. The loop checking coordinator makes sure that a list of the unit operations to be commissioned is communicated to all parties. Each of the unit operations on the list has a group of instruments that must be commissioned.

A Gantt chart may be used to lay out the order of commissioning if it is presented in adequate detail. At minimum, this chart helps to create a generalized commissioning strategy because it indicates which areas of the plant are to be ready and when. The Gantt chart in Figure 7-2 shows that Area 200 will be completed (electrically) by February 7. It will be followed by Area 300 dryer loops, which are to be completed by March 7. Because electrical connections must be completed before loop checking can occur, the Raw Materials Area is the first area to be loop checked.

The coordinator and all others involved are responsible for using safe practices at all times during this work. Because electrical circuits and other sources of energy are now being energized to verify that they work, proper lockout/tagout procedures must be followed when working on loops and on energized equipment.

Some companies prefer to use operators at the console to assist with calibration, loop checking, and function tests, other companies only use operators during loop checks. Depending on the company or complexity of the control schemes as well as how busy the operator is with other tasks, a controls systems engineer, a specialist, a contractor, or someone specifically trained on the PAS performs this task so that all

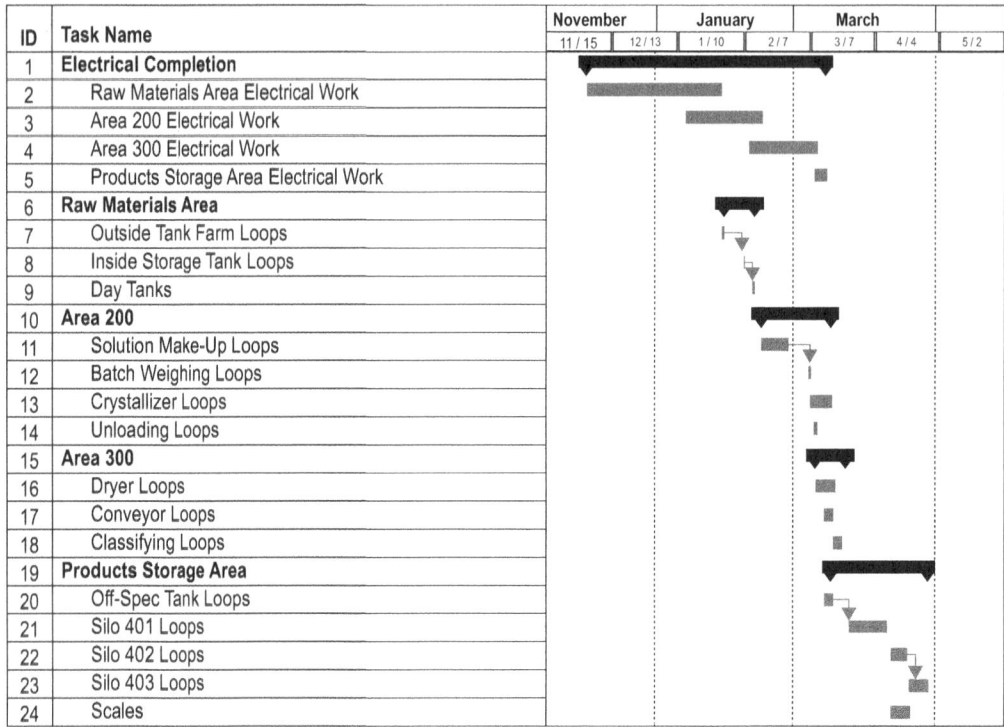

Figure 7-2. Example Gantt chart.

connections can be verified and any corrections to the configuration, graphics, and alarming are done at the same time. The loop is not accepted and handed over to Operations until it checks out completely. Any additional data or redlines developed during the test are kept in the loop folder.

There are ISA standards for loop sheets, but there is no standard type of document for performing loop and function checks. One methodology is using a list of equipment that must be checked out prior to plant start-up in the order dictated by the PM or start-up team. Some plants keep a single list of all loops to be checked out, arranged by tag name, while others have a stack of loop folders.

Figure 7-3 is an example of a loop check log sheet that has been partially completed by a person at the console working with a technician in the field during loop checks. If problems are encountered during loop checking, the loops are not checked off as complete. The start-up team decides whether to fix each problem as it is found or to get correctly functioning loops online first. Verification that loop checking is complete is then accomplished by looking at the loop check log sheet or a group of loop folders usually located at the PAS console.

Loop Check Log Sheet							
Project LC-1085							
Control Room							

Valves and Motors: Toggle/stroke output to each state (e.g., OPEN/CLOSED; START/STOP) and verify matching feedback.
*Transmitters and Control Valves: Simulate field signal (4–20 mA 5-point check).
Stroke valves and check valve position.

Tagname	Output (OP)	Field Signal* (PV)	Feedback (PV)	Notes (include problems encountered/solutions)	Initials	Date	OK
1. UV64555	OPEN/CLOSED	N/A	OPEN/CLOSED	Valve moves slowly	DRB	12/12/19	✓
2. FC64554	0–100%	0–250 SCFM	0–100%	N/A	DRB	12/13/19	✓
3. PC64118	0–100%	0–150 psig	0–100%	N/A	DRB	12.13.19	✓
4. PC64191							
5. LI12500							
6. LI13540							
7. FI15400							
8. FI1456705							
9. SI50402							
10. TC34659							
11. TI34321							
12. PDI86543							
13. VI23450							
14. LC63999							
15. DI53555							

REF: P&ID D-1715

Figure 7-3. Example of a loop check log sheet (in process).

Clearly, plants vary in the way they maintain and utilize their paperwork. There is no one set of standard procedures for loop checking, and plants with hazardous processes must take greater precautions than those without. In addition, the type of paperwork used varies from plant to plant, as does the methodology for verifying calibration and loop checking.

Relevant Standards

International Electrotechnical Commission

- IEC 60770-1, Ed. 2.0, *Transmitters for Use in Industrial Process Control Systems – Part 1: Methods for Performance Evaluation*, July 2010

- IEC 60770-2, Ed. 3.0, *Transmitters for Use in Industrial Process Control Systems – Part 2: Methods for Inspection and Routine Testing*, November 2010

- IEC 60770-3, Ed. 2.0, *Transmitters for Use in Industrial Process Control Systems – Part 3: Methods for Evaluation of Intelligent Transmitters*, May 2014

- IEC 61298-1, Ed. 2.0, *Process Measurement and Control Devices – General Methods and Procedures for Evaluating Performance – Part 1: General Considerations*, October 1, 2008

- IEC 61298-2, Ed. 2.0, *Process Measurement and Control Devices – General Methods and Procedures for Evaluating Performance – Part 2: Tests under Reference Conditions*, October 1, 2008
- IEC 61298-3, Ed. 2.0, *Process Measurement and Control Devices – General Methods and Procedures for Evaluating Performance – Part 3: Tests for the Effects of Influence Quantities*, October 1, 2008

Summary

During prestart-up, there is a lot of activity in an effort to reach the next phase of the project—*start-up*. All personnel involved with the project must keep safety tantamount to all else. Construction, inspection, configuration, and other pre-commissioning activities are occurring, sometimes simultaneously and in close proximity. All must be aware of SIMOPS situations to ensure safety is not compromised.

The CST uses tools and test equipment to diagnose and fix faulty loops and may be required to work with manufacturer's representatives and specialists to solve the plant's problems. The CST's experience and training will help to expedite these procedures and enable the plant to return to safe, optimum working order.

Review

7.1   Provide examples of how the Operations department can give you input toward solving a problem.

7.2   Name some items a quality assurance (QA)/quality control (QC) inspector might find that should be added to a punch list?

7.3   How do calibration and loop checking influence rework during commissioning and start-up, and how do they affect the timing of commissioning and start-up?

7.4   Why do vendors charge for calibration at the factory?

7.5   Why is it important to tag instruments as "calibrated" after they have been calibrated?

7.6   How does calibration methodology affect ISO 9000:2015 certification?

7.7   What standard document is referred to for engineering design, calibration, and loop and function checks?

7.8 Where might calibration and loop and function checks documents be available?

7.9 Why is it important that qualified personnel be involved in calibration and loop and function checking efforts?

7.10 What are some examples of reasons instruments may need to be recalibrated?

7.11 How do you think calibration documentation affects Occupational Safety and Health Administration (OSHA) process safety management (PSM) compliance?

7.12 Why is it important to have good communications skills when working on a problem during start-up?

7.13 Why might it be important to recheck instruments that were calibrated at the factory?

7.14 Name some tools that are used to troubleshoot plant problems. Explain how they are used.

7.15 What does *mechanically complete* (MC) mean? Why might a project not wait until everything is MC to move on to the next phase of the project? Why might this involve simultaneous operations (SIMOPS)?

7.16 What is the difference between pre-commissioning and commissioning?

7.17 What are some questions a CST may ask while he or she is checking a problem with a control loop?

7.18 What tool may be used to troubleshoot an analog transmitter?

7.19 Why is vendor documentation important when troubleshooting a problem?

7.20 Name some things that are completed during process automation system (PAS) configuration.

7.21 Why is it important to work with a vendor representative or equipment specialist?

7.22 Who are some people you can go to for assistance?

7.23 What is the difference between a safety instrumented system (SIS) and a safety instrumented function (SIF)?

7.24 What plant assistance might be required during start-up?

Recommended Reading

Cable, Mike. *Calibration: A Technician's Guide.* Research Triangle Park, NC: ISA (International Society of Automation), 2005.

Jeffery, Harley M. *Loop Checking: A Technician's Guide.* Research Triangle Park, NC: ISA (International Society of Automation).

Mostia, William L., Jr. *Troubleshooting: A Technician's Guide.* 2nd ed. Research Triangle Park, NC: ISA (International Society of Automation), 2006.

8

The Start-Up

Start-Up Overview

PSSR

Start-Up Plan

Issues and Problems

Final Acceptance Audit

Operation and Post-Project Activities

The time has come to start up the new part of the plant, the process section, and/or the entire (grassroots) plant. Management has approved the completed work and will soon close out the project. There are still quite a few activities that must to be completed before the plant can be started up.

8.1 Start-Up Overview

This author considers a project ready to be started up when all of the following are complete:

1. Construction has been deemed mechanically complete.

2. The process automation system (PAS)[1] has been installed, software is configured and validated, and personnel have been trained to use it.

---

1 Includes third-party and licensed systems and software.

3. Safety systems (e.g., SIS, fire and gas—F&G, deluge, and eyewash and safety showers) have been commissioned and are in working order.

4. Process lines are clean, blanks/blinds/locks have been removed, and lines/vessels/systems (when applicable, i.e., for processes with flammable and/or combustible materials) have been purged.

5. Management has approved completion, safety checks, and reviews (e.g., pre-startup safety review—PSSR), and there is a start-up plan in force that has been communicated to all parties.

All these items have been included in the project schedule (Gantt chart). The project manager (PM) and the project team have been monitoring the completion of these items and any additional costs associated with them as the project has proceeded.

The start-up team has been meeting for weeks, possibly months, preparing for the time to commence start-up. As with all the project activities thus far, safety is tantamount.

The multidisciplinary team (management, engineers, instrument and electrical (I&E) technicians, mechanics, etc.) has the appropriate documentation and work schedule to provide plant coverage around the clock (24/7) until management deems the plant is completely started up and producing product at appropriate levels (volume and quality).

There is most likely a procedure for starting up the plant, which includes the following: the areas that must be started up first, which vessels must be filled first, and what to do in case of an emergency or if bad-quality product is produced. This list and a what-if scenario are important as a guide to keep everyone working toward the same goal of a safe and efficient start-up with minimal problems. Of course, problems can occur, and it is everyone's responsibility to work to resolve them in a cooperative manner. Good recordkeeping and communication are extremely important because some events will be part of a steep learning curve for many participating in the start-up activities.

After weeks, months, and possibly years, the plant or new process section is ready to be started up. Depending on the type of project and facility, these types of tasks have been completed:

- Construction/installation/commissioning is complete for the following:
  - Infrastructure (civil; electrical grounding; utilities; fiber-optic network; power; buildings; and environmental, such as stacks and waste treatment)
  - Tie-ins to the existing facility, if applicable

- Equipment (vessels; motors; instrumentation; and safety, such as eyewashes and safety showers)
  - Pressure testing, dry/wet runs
  - Motor rotation checks—bump testing and lubrication
  - Specialized equipment testing with vendor support
- Instrumentation and associated electrical and pneumatic connections
  - Calibration, loop, and function checks completed
- Safety instrumented systems (SISs), PASs, subsystems, and interfaces installed, and the following associated software completed and tested:
  - I/O databases and control logic
  - Graphics
  - Historian and trends
  - Sequences
  - Licenses obtained
  - Miscellaneous software applications installed
  - Testing completed (e.g., factory acceptance testing—FAT, site acceptance testing—SAT, and site integration testing—SIT)

- Documentation for all systems is available and up-to-date.

- Operators are trained and personnel are well staffed for the upcoming start-up.

Despite this list, a few other things must occur before the plant can be brought online (also known as *put on-stream*), making the product(s) it was designed to produce. These may include the following:

- **Power** – Starters should be energized, fuses installed, wires terminated, and locks and tags removed.

- **Instrumentation** – Instrument isolation valves are open.

- **Process lines** – Blinds are removed and start-up strainers are installed as necessary. Purging vessels and lines prior to chemical introduction is complete.

- **Special equipment needs** – Pumps that require seal water or cooling should have these elements running. Equipment should be run-in longer than the bump tests done during commissioning to check vibration and other machinery monitoring protection system (MMPS) inputs.

- **Fire and gas protection systems** – Where applicable, systems are installed and checked out as working properly and drills have been performed, such as evacuation to muster points or shelter-in-place.

- **Plant assistance** – Make sure that the vendor/manufacturer or service representatives for certain pieces of equipment are present if they are required to be. Personnel from sister plants who have experience in the process or in start-up may also be present.

- **Operations personnel** – CSTs are typically assigned to assist the personnel when starting up. You and the plant operator should stay in direct communication during this fast-paced and unpredictable part of the job. Make sure your two-way radio is charged. All personnel are trained on their respective jobs and roles during start-up, and Operations personnel are trained on the use of the PAS.

- **Environmental approval** – Environmental permits are secured; after the plant is permitted to start up and is online, verify that it is running within its environmental limits. The CST and others can look for visible emissions (plume, tank overflow, etc.) in addition to assisting with the instrumentation for environmental monitoring.

- **Record keeping** – The latest documentation is on hand for all systems, loops, chemicals (MSDS), and procedures. For the purposes of communication, plant baselining, and documentation, it is important to keep good records during start-up. Many things will change as people learn how to run the new plant. Accidents (personnel, environmental, equipment) may occur, and the data and information gathered will help determine what happened and possibly prevent it from happening again. Keep a notebook for your start-up information and notes.

- **Safety checks** – Everyone must verify that all safety checks are complete: tags, locks, flange covers, and insulation are intact; rupture disks and pressure relief valves are installed; safety equipment is available; and personnel are trained in safety and emergency procedures. This task also includes making sure you have a readily available list of emergency contacts in case of environmental, health, or safety problems. As a start-up team member, you must ensure that systems can safely be started up.

- **Pre-startup safety review (PSSR) completed** – A safety review has been completed, all parties involved have approved and signed off on the review, and the process is ready to be started up (see the following section).

This is still an active time, and safety is foremost. Until the "go signal" has been given by management, the plant cannot be started up. This is a critical time when safety, environmental, monetary, and personnel concerns come together.

A successful start-up depends on the skill of the personnel, the quality of the construction, and the design of the process. These factors, along with good leadership and good work processes, will ensure a safe and successful start-up.

Tasks the team must remember when starting up a plant include the following:

- Open valves slowly to avoid "hammer"—applies to steam, water, and high-pressure lines.

- Monitor pressures to prevent relief valves from lifting and rupture disks from bursting.

- Monitor levels to avoid overflowing vessels or running them empty.

- Monitor equipment for issues such as unacceptable run conditions, failure, high amps, or blocked lines (deadheaded or cavitating pumps).

- Take samples to the lab to check product quality—more frequently in the beginning.

- Take readings of process conditions—more frequently in the beginning.

After a period of time—days, even months—the plant begins to produce the materials it was designed to produce. Amounts and quality are good. Depending on the size of the project, this period could be short (weeks) or very long (months to a year). Baseline data supporting performance guarantees will be gathered for the appropriate amount of time, and those responsible for these guarantees will be paid and released from the project, depending on the contracts and agreements made.

There may be breakdowns of equipment, resulting in total or partial plant shutdown. If the problem is instrumentation, the CST will be involved. If it is mechanical, someone else might be responsible. These shutdowns will result in yet another start-up of part or all of the plant. Now that the start-up is underway, a great deal of attention must be paid to how the project is proceeding and when people will get into a mode of normal plant operation. When the plant is at stable operation, maintenance personnel (i.e., CST, I&E, and mechanical) will not have to work around the clock but will need to support the plant as it will be supported for years to come or until the next project begins.

For loops that do not behave as well as they could, the tuning parameters can be more effectively changed now that the plant is running and the actual process materials are moving through the lines and equipment.

8.2 PSSR

A CST may also be asked to be a member of a PSSR team. This team aids in the transfer of projects from the commissioning stage described in Chapter 7 to the start-up stage by confirming that what was built and installed conforms to the original design and standard, and that no hazardous situations were created during construction and commissioning.

There is no standard PSSR methodology; the documents used to ensure that the process is ready and signed-off on differ by company. The goal of a PSSR is to ensure that the plant is ready to be started up safely.

According to 40 CFR § 68.77:

(a) The owner or operator shall perform a pre-startup safety review for new stationary sources and for modified stationary sources when the modification is significant enough to require a change in the process safety information.
(b) The pre-startup safety review shall confirm that prior to the introduction of regulated substances to a process:
   (1) Construction and equipment is in accordance with design specifications;
   (2) Safety, operating, maintenance, and emergency procedures are in place and are adequate;
   (3) For new stationary sources, a process hazard analysis has been performed and recommendations have been resolved or implemented before start-up; and modified stationary sources meet the requirements contained in management of change, § 68.75.
   (4) Training of each employee involved in operating a process has been completed.

Typically, representatives from various disciplines and managerial levels attend a PSSR meeting and review PSSR documents before signing off that the process is ready to be started up. The CST may be involved with this very important process prior to the start-up of the plant.

Figure 8-1 is an example of a PSSR checklist.

Pre-Startup Safety Review Checklist	
Title/Equipment:	
Project Number:	
Department/Area:	
Inspection Date:	

Signatures below indicate acceptance that the process is safe and satisfactory to start-up with the exceptions noted.

Engineering / Maintenance	Date
EH&S	Date
QA Group	Date
Manufacturing / Operations	Date
Project Engineer	Date
Process Engineering	Date

Checklist Item No.	Details (reference category/item no.)	Responsibility	Complete Sign & Date
Category A Action Items - *Items to be completed BEFORE authorization and start-up*			
1.			
2.			
3.			
4.			
5.			
Category B Action Items - *Items to be completed AFTER start-up*			
1.			
2.			
3.			
4.			
5.			
6.			
7.			
8.			
9.			
10.			
11.			
12.			
Sign below only when all punch list "before start-up" items are completed			
Authorized:	Plant Manager Signature:		Date

Figure 8-1. PSSR checklist example.

ITEM NO.	CATEGORY	Yes/No	N/A
1.0	**GENERAL SAFETY**		
1.1	All personnel have received training on equipment and standard operating procedures (SOPs)		
1.2	Is personal protective equipment (PPE) specified in the work processes and/or SOP?		
1.3	Has PPE been provided to company employees?		
1.4	Have personnel been trained in the use of the PPE and is the training documented?		
1.5	Are the following work permit procedures in place: confined space entry, lockout/tagout, hot work, high work		
1.6	Have fire protection systems (FPSs) been inspected by the insurance company?		
1.7	Have FPSs had acceptance testing completed and documented?		
1.8	Have FPSs been tested and is there an inspection program in place?		
1.9	Are fire extinguishers, first aid stations, and PPE requirements available and posted in all areas including control room, motor control center (MCC), and substations?		
2.0	**MACHINERY/EQUIPMENT SAFETY**		
2.1	Is all machinery/equipment installed so it is stable and secure?		
2.2	Has access to moving parts or danger zones been prevented (e.g., guards and barriers)?		
2.3	Have equipment risks been eliminated or are engineering controls utilized to minimize the risks?		
2.4	Is there safe access to the equipment?		
2.5	Is equipment identified with start/stop and emergency controls?		
2.6	Can equipment be securely isolated from ALL energy sources?		
3.0	**ERGONOMICS**		
3.1	Has equipment been designed to minimize stooping, bending, stretching, and over-reaching?		
3.2	Has the need to lift, carry, push, or pull heavy loads been minimized?		
3.3	Are monitors, controls, and START/STOP/ EMERGENCY buttons readily visible and accessible?		
3.4	Is glare minimized on operator workstation monitors?		
3.5	Can workstations be adjusted to a comfortable position?		
4.0	**OCCUPATIONAL HEALTH**		
4.1	Is respiratory protection specified?		
4.2	Is an occupational health monitoring program in place?		
4.3	Have SOPs been developed for any health hazards associated with this equipment?		
4.4	Has adequate ventilation been installed and is it on an inspection schedule?		
4.5	Are there inspection/cleaning ports on ductwork?		
4.6	Is a noise compliance plan in place?		
4.7	Is all insulation complete?		
5.0	**PROCESS SAFETY**		
5..1	Are Safety Data Sheets up-to-date and available?		
5..2	Has a chemical interaction matrix been prepared?		
5..3	Is the process design basis, control philosophy, and sequence of operations documented?		
5..4	Have the recommendations from safety reviews been implemented? Record any incomplete items.		
5..5	Are all pressure relief devices (PRD) and pressures shown on the P&IDs?		
5..6	Have the PRD calculations been provided?		
5.7	Do the PRD vent to safe locations?		
5.8	Are there isolation valves that, if closed, will inhibit the operation of PRD?		
5.9	If 5.8 is yes, has Operations established SOPs to ensure that isolation cannot inhibit operation of the PRD.		
5.10	Are all PRD included in the preventive maintenance program?		
6.0	**MANAGEMENT OF CHANGE (MOC)**		
6.1	Has a MOC procedure and document been approved?		
6.2	Are all action items arising from MOC initiated during the project complete?		
6.3	Have all changes made during construction been recorded and authorized?		
7.0	**PROCESS HAZARDS ANALYSIS (PHA)**		
7.1	Have project PHAs been approved and a final report prepared?		
7.2	Are all action items from the PHA complete?		

Figure 8-1. *Continued.*

ITEM NO.	CATEGORY	Yes/No	N/A
8.0	**QUALITY ASSURANCE**		
8.1	Have quality assurance inspection reports been completed and reports filed?		
8.2	List specific items field checked as part of this PSSR to ensure that:		
	The construction meets the design specifications		
	The construction matches the drawings		
8.3	Have the following documented been provided and approved?		
	Instrument indexes and instrument loop diagrams		
	Interlocks and trips (hardwire and software), process alarms, and permissives		
	As-built drawings (P&IDs, electrical, piping, and mechanical)		
	Data sheets, loop diagrams		
	Non-destructive test (NDT) certifications		
	Electrical certification for classified areas		
9.0	**MECHANICAL INTEGRITY**		
9.1	Have maintenance procedures been approved?		
9.2	Have maintenance personnel been trained?		
9.3	Is there a spare parts list and a parts ordering system?		
9.4	Is there an adequate supply of spare parts, operating supplies, and maintenance materials?		
9.5	Are inspections and tests for the following equipment been included in a PM schedule?		
	Pressure vessels and storage tanks		
	Pressure relief systems, vent systems, and devices		
	Critical controls, interlocks, alarms, and instruments		
	Emergency devices (including shutdown systems and isolation systems)		
	Fire protection equipment		
	Piping systems (including valves, flanges, and expansion bellows) in critical service		
	Existing process tie-ins		
	MCC starters		
	Emergency alarm and communication systems		
	Instrumentation		
	Pumps and other rotating equipment		
9.6	Is reliability engineering analysis complete for process safety management (PSM) critical equipment?		
9.7	Is the equipment inspected with certificates (by outside personnel) on file?		
9.8	Have all commissioning tests or inspections been completed?		
10.0	**OPERATING PROCEDURES AND SAFE WORK PRACTICES**		
10.1	Have SOPs been prepared, updated, and approved?		
	Do the procedures cover:		
	Initial start-up		
	Normal start-up		
	Normal operations		
	Normal shutdowns		
	Emergency shutdowns (ESD)		
	Start-up after ESD		
	Start-up following turnarounds		
	Nonroutine procedures (e.g., equipment clean-out and equipment preparation for maintenance)		
	Auxiliary equipment operations		
	Safety and operational issues		
	Change control procedures		
11.0	**TRAINING AND PERFORMANCE**		
11.1	Has specific process (or job task) training been given to personnel?		
11.2.	Have training records been updated?		
12.0	**CONTRACTOR SAFETY**		
12.1	Are all contract personnel trained in: e.g., chemical awareness, operating activities, and evacuation procedures?		

Figure 8-1. *Continued.*

ITEM NO.	CATEGORY	Yes/No	N/A
13.1	Have the alarms/interlocks been classified and designed by the project team?		
13.2	Did the loop testing confirm that the actions proved to be fail-safe?		
13.3	Has an interlock/critical alarm SOP for testing, through to the final element, been prepared and reviewed for each control system?		
13.4	For alarms/interlocks, have all possible interlock routes been tested?		
13.5	Has all documentation been updated (e.g., interlock lists, P&IDs, and logic drawings)?		
13.6	Does the control system documentation adequately specify the following:		
	All major components and their model and serial numbers?		
	All communication cables layout and configuration?		
	Any configurable or custom settings and set-up?		
13.7	Are fire detection and prevention systems tested and online?		
13.8	Are procedures for software protection developed and available to personnel?		
13.9	Is the software properly documented and filed (e.g., logic drawings, schematics, and sequence/batch descriptions)?		
13.10	Has all software been validated and tested?		
13.11	Is there verification that the equipment does not restart, either on the resetting of a protective device, such as an interlock, or the reestablishment of power after an outage?		
14.0	**ENVIRONMENTAL**		
14.1	Are all secondary containment/bonding facilities adequate?		
14.2	Have arrangements been made for the ID, classification, and disposal of waste materials?		
14.3	Have all materials used in the system been entered into a chemicals inventory list?		
14.4	Are updated spill SOPs available?		
14.5	Are material loading/unloading facilities constructed in accordance with corporate EH&S standards?		
14.6	Is there adequate containment (110% of truck volume) in the unloading areas for bulk liquid chemicals?		
14.7	Have the corporate EH&S guidelines been followed during the design stage of this project?		
14.8	Have all waste streams been identified, quantified, analyzed, and minimized?		
14.9	Are all of the applicable environmental and operating permits in place?		
15.0	**EMERGENCY RESPONSE**		
15.1	Have all necessary precautions been taken to ensure that the equipment is not a source of ignition to any flammable materials, irrespective of their source?		
15.2	Are fire protection facilities adequate?		
15.3	Are emergency escape routes adequate and properly indicated?		
15.4	Is emergency lighting adequate?		
15.5	Have emergency procedures been prepared and personnel trained?		
15.6	Is the community panel advised of proposed new major projects?		
15.7	Has an electrical safety checklist been completed?		
15.8	Has the equipment been properly installed and constructed to corporate guidelines and local legislation, including any special installation requirements?		
15.9	Has equipment been designed and purchased for the conditions under which it will operate (e.g., hazardous areas)?		
15.10	Are all live parts adequately enclosed to prevent access?		
15.11	Does grounding and bonding comply with corporate and local standards?		
15.12	Have fuses or circuit breakers been provided to automatically disconnect the supply?		
15.13	Is all documentation and drawings "as-built"?		
16.0	**FIELD VERIFICATION**		
16.1	Is the normal lighting adequate for normal and maintenance operations?		
16.2	Is emergency lighting sufficient?		
16.3	Are all extreme hot and cold surfaces in the proximity of personnel insulated?		
16.4	Are all vessels, instruments, equipment, and piping adequately labeled?		
16.5	Is there any rusted and/or damaged equipment?		
16.6	Are swing gates or chains installed at the top of ladders and/or on access platforms?		
16.7	Are there any gaps between platforms and equipment that could create a foot hazard?		
16.8	Is equipment and platform access adequate?		
16.9	Are all electrical switches, disconnects, MCCs, control panels, cables, etc. labeled?		
16.10	Are there provisions for electrical and/or mechanical isolation of equipment?		

Figure 8-1. *Continued.*

ITEM NO.	CATEGORY	Yes/No	N/A
16.11	Are points of isolation clearly marked and readily accessible?		
16.12	Are wall penetrations adequately sealed?		
16.13	Are electrical conduits sealed in accordance with code requirements?		
16.14	Are evacuation routes clearly marked?		
16.15	Are fire extinguishers installed properly?		
16.16	Are emergency stops provided where there is a potential for entrapment or exposure?		
16.17	Has all scaffolding and construction equipment been removed?		
16.18	Is housekeeping acceptable?		
16.19	Are safety showers and eyewashes provided and adequately marked?		
16.20	Are safety showers and eyewashes routinely inspected?		
16.21	Are safety showers and eyewashes easily visible and accessible?		
16.22	Is there sufficient lighting everywhere?		
16.23	Are all overhead fixtures properly secured?		
16.24	Have noise-monitoring evaluations been completed?		
16.25	Have signs been posted where noise levels excess 85 dB?		
16.26	Are earplugs available near areas exceeding 85 dB?		

Figure 8-1. *Continued.*

8.3 Start-Up Plan

Many factors determine how a plant is started-up. The start-up plan for the project will involve the following, at minimum:

- **Staffing** – This includes division of responsibility, task assignments and responsibilities, scheduling, workloads and priorities, coverage during different start-up phases, overtime, and training.

- **System start-up order** – The order depends on whether the project is a grassroots plant or additions to an existing plant.

- **Introduction of feedstock** – This is a critical part of the plan as flammable, combustible, or dangerous chemicals must be introduced into the new process with extreme care. If there is a leak that needs to be repaired after chemicals are introduced, then an intricate procedure of hazardous chemical removal, including cleaning and purging, will be necessary before the repair can be made.

- **Initial ramp-up** – Plans to ramp up the plant from completely down to full capacity must be carefully thought out. Personnel are going through a steep learning curve, equipment may be heating up, and a minimum of off-spec product should be produced, otherwise considerations for its removal and/or rework must be made.

- **ESD testing** – A start-up plan should involve emergency shutdown systems testing. This is critical as the ESD system must work once the new process is online. This testing normally occurs before the plant is running to ensure that it works properly when it is needed. If the ESD system fails to work as designed, repairs or changes can be completed before the plant is started up, avoiding the need to shut the plant down to make the necessary changes.

- **Loop tuning** – The process should be online for loop tuning to be attempted because it is unrealistic to think that a loop will perform the same offline as it will online. As the process is started up, different operating conditions occur, therefore loop tuning is often an ongoing work process to keep the plant in control and running as optimally as possible until it is at steady-state.

- **Environmental testing** – The process should be online and at levels where environmental permits were obtained. Therefore, testing will occur as the plant is starting up and when it has attained the permit levels. The start-up plant should include provisions for maintaining operating conditions at levels to permit proper testing for internal and external authorities.

Each of these items will be discussed in greater detail next.

8.3.1 Staffing

Project start-ups can run from a few days to a few weeks to over a year, depending on the size, scope of the project, and the interactions between different parts of the process. Staggered start-ups of different parts of the plant often occur when there are duplicate trains producing the same product. Rarely does the entire (pertaining to grassroots) plant/project start up at one time.

The management of project phases requires careful planning of staffing (people, by discipline and level of expertise), materials, and the product production schedule. The project Gantt chart (see Chapter 3) is a living document that reflects project progress and is modified as work is accomplished; it highlights the work yet to be completed. Management looks at this chart and the production schedule (which usually includes future orders) to determine how to staff the plant to complete installation and start-up on time and within budget. A start-up checklist and plan will most likely be developed as the time for start-up approaches. These will detail activities and the order in which they must be accomplished to safely and effectively start up the plant.

Plant management is a group of managers from several departments that usually includes the plant manager (typically the highest plant position), operations manager, technical manager (process engineering), project engineering manager, sales manager, and the PM for the project. The start-up order is agreed on by this group of managers and is reflected in the Gantt chart (project schedule). This group of managers has knowledge about the requirements of the plant (e.g., sales orders) and is responsible for meeting the project schedule; ensuring the quality of product produced from the new plant; and meeting all safety, environmental, and health requirements for running a production facility.

The I&E shop and the CST are part of a plant's service organization. That is, they must help Operations as needed before and during start-up and when the plant is finally online. The plant's safe and successful operation is the number one priority. The CST can be called on to help by phone (either emergency or planned) or by written work order. In a unionized plant, a CST will be called according to the union contract rules.

If the call is an emergency, the CST must be able to respond quickly. This may require that you come in from home or off-site. Once at the site, you must communicate with the operator about the problem, identify the problem from the operator's explanation and from observed symptoms, and then correct it.

You must also decide when to call for additional assistance from specialists or vendor representatives, other CSTs, I&E technicians, control systems engineers, system integrators, or other engineers.

When you respond to a planned call or work order, the steps are the same. The difference is that in an emergency, decisions must be made quickly and there is little time for mistakes.

Much of how a CST supports Operations and other site personnel must come from training, individual study, and years of working in a plant.

Starting up a process, whether continuous or batch, requires the careful coordination of many variables, both process and administrative.

8.3.2 System Start-up Order

Examples of how the completion order might be determined and/or affected follow:

- If this is a grassroots plant, then obviously utilities and infrastructure must be started up first. This is so that systems that provide things like steam, plant air, and process water will be ready for plant use.

- The utility section or raw materials section (upstream) of a plant must be started up before the other downstream plant process sections because of the dependence of the downstream processes on the upstream plant sections. Likewise, downstream tanks and storage must be ready before the process starts up because these vessels are needed for product and off-spec materials.

- A piece of equipment (e.g., distillation column, Distillation Column B – Train 2) is being added to an existing plant with a (same design) distillation column, Distillation Column A, which is already running (Train 1). There are common line tie-ins (e.g., feed and electricity) between Column A and Column B. It might not be possible to complete the tie-in and therefore start up Column B until Column A is down for some reason (e.g., scheduled maintenance).

- A flammable liquid might not be introduced into the part of the plant that requires it until the rest of the process is checked out and running. This avoids the delays that would otherwise arise if problems that require welding arise, in which case the area would have to be completely washed out and tested for the presence of flammable or explosive vapors before proceeding.

- If unexpected delays occur—for example, a major piece of equipment has not been installed because delivery was delayed—the project schedule will have to be modified.

- Environmental permitting or stack testing must occur before the entire process can be started up.

8.3.3 Introducing Feedstock

If feedstock is flammable, combustible, or inherently dangerous, then vessels and pipelines must be free of oxygen. Normally nitrogen blanketing or purging is utilized before the feedstock flows into the new plant.

Most processes require totally clean lines and vessels; therefore, cleaning and start-up screens and filters are employed. After a period of time, the start-up screens and filters are removed or replaced and now contain chemicals that must be handled carefully. If something goes wrong (e.g., equipment failure), the process may need to be totally shut down, feedstock isolated, and vessels may need to be opened and totally cleaned out before entry is possible.

Now that feedstock has been introduced, reactions occur and process variables (PVs; e.g., temperature, pressure) are changing (high/low). Some people may not be as experienced as others who need to run and monitor the new process. When feedstock is introduced into the new process, extreme care must be taken as the start-up team is now under a lot of pressure and situations may be getting more hazardous.

8.3.4 Initial Ramp-up

The following activities are generally completed during the project start-up phase:

- **Power** – Starters should be energized, fuses installed, wires terminated, and locks and tags removed.

- **Instrumentation** – Instrument isolation valves are open.

- **Process lines** – Blinds are removed and start-up strainers are installed as necessary.

- **Special equipment needs** – Pumps that require seal water or cooling should have these elements running.

The items in this list will be explored in greater detail in the following section. When a process comes online, it is normally brought up slowly. As mentioned earlier, PVs are rapidly changing, a lot of control points must be monitored, alarms may be flooding in on the PAS, and loops may need to be tuned. Depending on how complicated the process is, automation sequences may not be run automatically initially, so additional operator monitoring and intervention is required to control parts of the process that will be controlled automatically in the future. Oftentimes, plants man the initial start-up phase with more people than will be running the plant in the future when the plant is at *steady-state*.

This is a hectic time and there will be probably be a lot of people observing what is going on. The control room may be fairly crowded with personnel from various departments, including but not limited to: operators, management, I&E, CST, PAS personnel, engineers, and vendors. Not everything will go smoothly and there will be a lot of opinions regarding what must be done. Troubleshooting will be required to solve problems as quickly as possible. It is important for the team to work together to ensure the project is started up safely and it will eventually make the desired product at the quality specifications it was designed for.

8.3.5 Emergency Shutdown Testing

Depending on the process, it is important for the emergency shutdown (ESD) system to be tested before starting up. Because multiple final elements (e.g., valves and motors) must move to their fail-safe position quickly at the push of a button or "trip," these devices must have been loop and function checked to ensure they will work when required. If not, those that do not work will need to be rewired, repaired, or replaced. This is the time to ensure that the system works properly, before feedstock is introduced and/or the plant is started up.

In addition to the ESD system testing, other ESD procedures should be practiced, for example, having people gather at safe (muster) points who know how to use safety equipment and who to contact in case of emergency.

Finally, eyewash and safety showers, fire and gas detection, and deluge systems should be tested at this time if they are part of the process.

8.3.6 Loop Tuning

The person (or persons) responsible for performing the configuration of the PAS (distributed control system—DCS, basic process control system—BPCS, and programmable logic controller—PLC) database often start(s) with the best approximation of tuning parameters or initial settings for each type of control loop (e.g., proportional-integral-derivative—PID) before the plant starts up. Once the plant is running, these parameters need to be fine-tuned by watching the process response. This is a time-consuming but important process for safety, efficiency, and product quality control purposes. This is not easy because the plant may not be at full capacity, and it is probably not at steady-state (is up and down), and some people are somewhat inexperienced in running the plant and/or performing loop tuning.

Different plants assign responsibility for the tuning operation to different people. Often a CST is involved with tuning, if not entirely responsible for it. The CST may also work with a control systems engineer to perform or learn about the loop tuning process. Plants utilize experienced personnel who know how certain types of loops react, what tuning parameters result in fast or slow response, and the type of process for which the loops are being used. They also know how to use tools, including simulation software packages, to help make the tuning process more efficient.

Loop tuning can be a tedious chore that must be repeated for many loops in the plant. Tuning a PID controller entails selecting the right combination of proportional, integral, and derivative action to achieve the desired process control. It depends on the process, its response to changes, and the availability of time and opportunity to make changes to the process. It is not advisable to "bump" the process through loop tuning efforts. Finally, it also depends on the type of controller and its associated control algorithm which is control system dependent.

Loop tuning software (analytical software solutions) can speed up the loop tuning process and make it more scientific and reliable. This software captures important data, such as the loop's historical performance, along with information like real time and historical trends of the PV, set point (SP), output, error, and tuning parameters. It can even act as a simulator or be part of a simulation software package.

Many plant projects purchase process simulators and include the activities of modeling and process simulation in the schedule (shown as activities on the Gantt chant). *Process modeling* is a computer modeling technique that typically involves using software to define a system of interconnected components of the real process, which are analyzed so that the steady-state or dynamic behavior of the system can be predicted through process simulation. *Process simulation* can be used to design, develop, analyze, and optimize a process. Depending on the size of the project, process simulation can initially be considered expensive, but its use results in "payback" in many forms by providing the ability to quantify system performance by the following means:

- Throughput under average and peak loads
- System cycle time (how long it takes to produce one part)
- Utilization of resources, labor, and machines
- Bottlenecks and choke points
- Queuing at work locations
- Queuing and delays caused by material handling devices and systems
- Work in process storage needs
- Staffing requirements
- Effectiveness of scheduling systems
- Effectiveness of the PAS
- Training

If a simulation program is not used, the process has to be running to tune its control loops. Tuning most often occurs *after* the start-up is underway because the plant is running closer to steady-state, with real process materials in the lines.

Care, knowledge of the process, and communication with operating personnel are imperative while making any process changes, including loop tuning. Before making a tuning change, take the following steps:

1. Ascertain whether an MOC form or procedure must be initiated. Complete this paperwork if necessary.
2. Determine if the process is running at steady-state.

3. Make sure no operating changes are imminent.

4. Back up the software and controller settings that you will be working with before you make changes. Make hard copies if desired.

5. Notify Operations that you will be performing loop tuning.

6. Write down the changes as you make them.

7. Wait a sufficient amount of time between each change before making another change. Once you are satisfied with the final change, wait a little longer and leave word (oral and written) about what you did.

8. Make a backup (not in the same file as the first) of the new configuration.

These steps will help you safely make loop tuning changes, as well as document the original and the new parameters. If you must revert to the original configuration, you will have records and a backup of this information.

Some loops cannot be taken off-line once the plant is running. This is because they are connected to, for example, a loop involved with an interlock or an SIS. If these critical loops are taken off-line or their alarm or shutdown limits are met (either by a real PV or by someone simulating a signal), they can shut down the plant or cause an unsafe or unexpected situation. It is important to know what each loop is used for and how to work with it. This information should be available from documentation, plant personnel, and sometimes field markings (tags or paint colors). You must communicate with all personnel affected by each loop you work with so that accidents or unsafe conditions are averted.

This is a very basic loop tuning discussion. Refer to the "Recommended Reading" section at the end of this chapter for further information.

8.3.7 Environmental Testing

When the plant is running well, baseline samples must be taken to ensure that plant effluents (air, water, land) meet permit requirements. The results of these samples may also indicate that instrumentation or equipment is not working properly and a vendor representative needs to assist with solving the problem. Additionally, if the plant is not running well, samples taken during these process conditions will indicate there is a problem and whether an environmental agency notification is required. Conversely, results indicating that the plant is running well and that a performance milestone has been successfully met or that a piece of equipment has met the performance specifications may allow final payment to a supplier.

8.4 Issues and Problems

Other things that may hinder start-up and thus affect staffing (i.e., manning) are environmental, processing, and equipment problems.

8.4.1 Environmental Problems

Once a plant goes online, samples may need to be taken of effluents to determine whether permitted air, ground, and water levels are being met. If the samples show this is not the case, the plant may need to be shut down and modifications made. See Chapter 1 and Chapter 4 for more information on environmental requirements and how the CST is involved with environmental instrumentation and monitoring systems as they apply to the start-up.

8.4.2 Processing Problems

An example of a processing problem is material that does not flow well from vessels. Again, in such a situation the plant may have to be shut down and modifications made to correct the problem. Typically, the CST will use tuning or functional testing to resolve problems if instrumentation and control have either caused or can help alleviate the problem. In this case, the CST will work closely with plant and engineering personnel to determine whether he or she can help.

8.4.3 Equipment Problems

If new equipment breaks down or fails to perform as expected, it must be replaced or modified by plant personnel. A service or manufacturer's representative and/or contractors may also be called in to look at the equipment, which can take anywhere from a few days to several weeks. During this time, you may be asked to work on other things or be assigned to expedite problem-solving with the service representative or contractors.

If equipment must be replaced, you will be required to assist in disconnecting the equipment and its associated instrumentation. When the new equipment and instrumentation are installed, you will help with connection, calibration, and loop checking. Calibration and loop checking are necessary to ensure that the new equipment and instrumentation are working properly and are once again communicating with the PAS.

8.5 Final Acceptance Audit

Other tests that the company may require are performed after the plant has started up. These tests require their own set of documentation, and the CST may be involved in such testing. During plant design, many documents are prepared to ensure that the plant meets the criteria desired by the plant operating company to make the product according to specification or design. One of these documents is a functional specification, discussed in Chapter 3.

Depending on the process, functional specifications, licensor requirements, contracts or agreements, and management requirements, performance guarantees, benchmarks, key performance indicators, and/or beneficial operation may be milestones that must be achieved. The tests may be a condition of payment, which means they are performed before the customer finally accepts the system and enters into a maintenance agreement with the supplier of the system or licenser of the technology. Other parties may now be allowed to leave the plant, their job is done, and others, including engineers and CST, may now be able to return from working around the clock, on shift, to fewer hours during normal hours (e.g., 9 to 5).

For performance to be measured, data must be collected and samples of the intermediary and finished products taken and analyzed. If a transmitter is malfunctioning, the data for these measurements will be compromised, so a CST may be engaged to help. Because this testing normally takes place over a predetermined amount of time, time is of the essence. While fixing this loop as quickly as possible is important, safely should remain paramount.

A preliminary test report or letter is generated to allow the plant management to declare "substantial completion" or "beneficial operation," and commence commercial operation. As with factory acceptance, site integration, and site acceptance testing, availability and performance testing requires the involvement of the appropriate personnel to plan and execute these tests.

8.6 Operation and Post-Project Activities

8.6.1 Sustainable Steady-State Operation

It is everyone's hope that soon after the plant start-up the process achieves *steady-state*. A system or a process is at steady-state if the variables that define the behavior of the system or the process are unchanging[2] in time.

Examples of indications that the process is at steady-state are:

- Stable process conditions as indicated by process trends
- Consistent samples from various intermediate process locations
- Consistent product as analyzed by the laboratory or continuous samples

---

[2] Unchanging does not necessarily mean constant, but within an acceptable dead band (plus or minus) or range, as defined by a process authority.

Although steady-state operation occurs for an extended period of time, fluctuations, for various reasons, require the PAS or Operations and Maintenance personnel to make changes to get the process "back in control." Some examples of why the process may cycle a small amount or to a large extent are the following:

- Variations in feedstock
- Loops that need to be tuned better
- Equipment irregularities or failures (e.g., valves and transmitters)

8.6.2 Project Close-Out

Other activities that often occur during project close-out are:

- Contracts are established with some vendors for future support of the operating systems. Examples of these types of contracts are:
    - 24/7 support for the PAS using the vendor's Technical Assistance Centers (TAC). Normally an 800 number is provided to call for Q&A about the system, call-outs for on-site maintenance, and, where applicable, remote dial-in support.
    - Spare parts (onsite or offsite) for various systems.
    - Documentation and software updates provided on an annual basis.
    - Reduced rates for training courses provided by various vendors.
- Documentation is updated to reflect the "as-built" status of the project. The work to accomplish this may be done in-house (by plant personnel) or contracted out to another company or the engineering firm that originally designed the project. Redlines marked by all involved with the project from the beginning should be provided to whomever is making these updates so that all changes can be captured to reflect the current plant design.
- If a complete maintenance work process has not been fully established during the project, one should be completed during this time. Elements of this system include:
    - Work order system for scheduling planned and unplanned (emergency) maintenance
    - Preventative maintenance program for all systems (mechanical, electrical, instrument, PAS)
    - Warehouse program for spare parts and equipment

- Lessons learned should be reviewed. These are experiences distilled from a project that should be actively taken into account in future projects.

    o These experiences are often captured at post-project meetings led by a facilitator familiar with the lessons learned process. The meetings should allow for free-flowing, uninhibited conversations describing both the good and bad experiences various disciplines encountered during the project. The meeting may be an unstructured "brainstorming" session or use a more formalized methodology. Regardless, lessons learned should be productive and not a "complaint session" but may include technical as well as safety (near-miss) topics. Lessons learned is an important process that should allow all those involved with the project to vent as well as teach others what went well and what could have been done better. CSTs can contribute greatly during these meetings, because they have worked on different aspects of the project and therefore have had various experiences.

Summary

Many things have to be completed before the start-up can commence, much having to do with maintaining a safe working environment by conducting a multi-disciplinary PSSR and informing all parties what the start-up plan is. Although issues and problems will occur, eventually the new process will be online running at steady-state, making the high quality product at the production levels it was intended to meet. The final acceptance audit will reward those who were involved with the project both monetarily and by recognizing their hard work. Even though the project is closed-out there will continue to be a lot of work for personnel at the facility maintaining equipment, instrumentation, and the PAS and getting ready for the next expansion project for this facility and subsequent start-up to make even more product.

Conclusion

A start-up is an exciting and challenging event. It is a time to learn and help others learn how to work together to bring a new plant or process online safely and efficiently.

Postscript: ISA CAP

ISA currently has two certification programs: Certified Automation Professional (CAP) and Certified Control Systems Technician (CCST). Now that you have gone this far in your studies, you may want to consider the ISA CAP Certification Program. The CAP is responsible for the direction, definition, design, development/application, deployment, documentation, and support of systems, software, and equipment used in control systems, manufacturing information systems, systems integration, and operational consulting. CAPs are automation professionals who have proven they possess an extensive knowledge of automation and controls. After passing the CAP exam, a CAP has documented evidence that he or she possesses the expertise and qualifications to excel in his or her field.

Review

8.1 Name some safety checks that must occur before the start-up of any equipment or system.

8.2 Why does a start-up team meet more frequently as the start-up time approaches?

8.3 Give an example of a processing problem.

8.4 What is meant by *troubleshooting*, and what are some of the tasks involved?

8.5 Name some environmental checks done while completing a pre-startup safety review (PSSR).

8.6 What things might hinder start-up and thus affect staffing?

8.7 What is a *performance guarantee*?

8.8 Why is personal protective equipment (PPE) an important part of a PSSR? What other items, for example, process area checks, are important to review during a PSSR?

8.9 What may a CST do during configuration?

8.10 What are some examples of special equipment tasks performed during start-up?

8.11 What is meant by *steady-state operation*?

8.12 What factors should be considered in connection with power in preparing for start-up?

8.13 Name some post-project activities that might involve a CST.

8.14 Name some factors that might influence the order of start-up.

Recommended Reading

McMillan, Gregory K. *Good Tuning: A Pocket Guide*. 4th ed. Research Triangle Park, NC: ISA (International Society of Automation), 2015.

Appendix A: Instrument Repair Technician Job Description

The responsibilities include the installation, troubleshooting, repair, and routine maintenance of instruments throughout the plant. A large portion of the work includes troubleshooting and coordinating repairs to working loops in the various operating units. The job requires the technician to follow company and craft standards and specifications, and coordinate all activities with Operations and with other crafts when performing various routine maintenance functions.

A.1 Essential Functions

Main duty areas and examples of essential tasks for each area:

A.1.1 Routine Instrumentation Calibration for the Following during Major Projects and Turnarounds

- Flowmeters
- pH meter; temperature, pressure and analyzer transmitters
- Pneumatic valve actuators
- Valve range control valves
- Valve positioners

A.1.2 Typical General and Preventative Maintenance Tasks

- Remove, replace, and/or repair old or broken flow, temperature, pressure, and level meters; electric and pneumatic transmitters and associated modules; D printed circuit boards; valve parts; transmitter wiring and modules; chromatographs; pH analyzers; transducers; and electronic and pneumatic valves.

- Ensure the control system database configuration is ranged correctly and aligned with instruments and field equipment where necessary.

A.1.3 Examples of Required Calculations

- Calculate voltages, amperages, resistance, pressure, etc.
- Configure circuits in series and parallel.
- Size control valves.

A.1.4 Troubleshooting Loops

- Check the wiring to and from the instrument for proper hook-up and physical condition.
- Use the proper simulator to check the output of the instrument, making sure that proper signal conversions are being received/transmitted.
- Check for a malfunctioning sensing element and replace the sensor if defective.
- Check for proper power to the transmitter.
- Check zero and span and adjustment as necessary.
- Stroke the valve using proper equipment for the initial check of the valve stem, actuator, and positioner.
- Check the actuator air supply.

A.1.5 Newly Installed Equipment Inspection

- Check the wiring using a voltage-ohm meter (VOM).
- Perform instrument calibration using a universal calibrator.
- Use the control system or asset management system to upload/download instrument ranges and check for irregularities.

A.1.6 Maintenance and Repair of Alarm Devices
- Adjust and troubleshoot temperature, pressure, flow, and level switches.
- Troubleshoot interferences to alarm devices.
- Function test, calibrate, and repair alarm and shutdown systems using diagrams and schematics.

A.1.7 Behaviors Required to Perform the above Essential Functions
- Safety awareness
- Technical judgment
- Conscientiousness
- Initiative
- Adherence to policies and procedures
- Maturity
- Teamwork and interpersonal skills

A.1.8 Physical Requirements
- Walking/standing/bending
- Climbing ladders (up to 100 ft or 30.48 m)
- Climbing stairs (over 100 ft or 30.48 m)
- Stepping over and crawling under structures
- Entering enclosed spaces
- Entering "tight" places where free movement is restricted
- Carrying objects (over 100 yd or 91.44 m)
- Working in contorted or awkward positions
- Working at tasks that require:
  - Muscular endurance and stamina
  - Good balance
  - Coordinated hand/arm/body movements

- Wrist and steady hand/arm movements
- Rapid eye-hand or eye-foot coordination
- Small, accurate movements of fingers
- Ability to see fine details
- Ability to hear and distinguish between different sounds
- Lifting, pushing, pulling, or dragging objects (over 50 lb or 22.5 kg)

A.1.9 Working Conditions
- Working in a noisy environment
- Working in extreme temperatures

Appendix B: Safety Data Sheet (SDS)

Safety Data Sheet (SDS)
OSHA HazCom Standard 29 CFR 1910.1200(g) and GHS Rev 03.

Issue date 11/22/2019   Reviewed on 11/22/2019

1 Identification

· **Product Identifier**

· **Trade Name:** Potassium Hydroxide, 40% Liquid
· **CAS Number:** 1310-58-3
· **Relevant identified uses of the substance or mixture and uses advised against:**
Source of Potassium Hydroxide
· **Product Description:** KOH 40% Liquid

· **Details of the Supplier of the Safety Data Sheet:**
· **Manufacturer/Supplier:**
NuGeneration Technologies, LLC (dba NuGenTec)
1155 Park Avenue, Emeryville, CA 94608
salesteam@nugentec.com   www.nugentec.com
1-888-996-8436 or 1-707-820-4080 for product information
· **Emergency telephone number:**
PERS Emergency Response: US and Canada - 1-800-633-8253, International 1-801-629-0667

2 Hazard(s) Identification

· **Classification of the substance or mixture:**

 Corrosion

Met. Corr.1   H290 May be corrosive to metals.
Skin Corr. 1A H314 Causes severe skin burns and eye damage.
Eye Dam. 1   H318 Causes serious eye damage.

Acute Tox. 4   H302 Harmful if swallowed.

· **Label elements:**
· **Hazard pictograms:**

· **Signal word:** Danger

· **Hazard-determining components of labeling:**
Potassium Hydroxide
· **Hazard statements:**
H290 May be corrosive to metals.
H302 Harmful if swallowed.
H314 Causes severe skin burns and eye damage.
· **Precautionary statements:**
P234   Keep only in original container.
P260   Do not breathe dusts or mists.
P264   Wash thoroughly after handling.

(Continued)
US

Safety Data Sheet (SDS)
OSHA HazCom Standard 29 CFR 1910.1200(g) and GHS Rev 03.

Issue date 11/22/2019   Reviewed on 11/22/2019

Trade Name: Potassium Hydroxide, 40% Liquid

P270	Do not eat, drink or smoke when using this product.
P280	Wear protective gloves/protective clothing/eye protection/face protection.
P301+P312	If swallowed: Call a poison center/doctor if you feel unwell.
P301+P330+P331	If swallowed: Rinse mouth. Do NOT induce vomiting.
P303+P361+P353	If on skin (or hair): Take off immediately all contaminated clothing. Rinse skin with water/shower.
P304+P340	IF INHALED: Remove person to fresh air and keep comfortable for breathing.
P305+P351+P338	If in eyes: Rinse cautiously with water for several minutes. Remove contact lenses, if present and easy to do. Continue rinsing.
P310	Immediately call a poison center/doctor.
P321	Specific treatment (see supplementary first aid instructions on this Safety Data Sheet).
P363	Wash contaminated clothing before reuse.
P390	Absorb spillage to prevent material damage.
P405	Store locked up.
P406	Store in corrosive resistant container with a resistant inner liner.
P501	Dispose of contents/container in accordance with local/regional/national/international regulations.

· **Unknown acute toxicity:**
0 % of the mixture consists of component(s) of unknown toxicity.
· **Classification system:**
· **NFPA ratings (scale 0 - 4)**

Health = 3
Fire = 0
Reactivity = 0

· **HMIS-ratings (scale 0 - 4)**

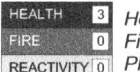
Health = 3
Fire = 0
Physical Hazard = 0

· **Hazard(s) not otherwise classified (HNOC):** None known

3 Composition/Information on Ingredients

· **Non-hazardous components:**

7732-18-5	Water, distilled water, deionized water	40-60%

· **Chemical characterization: Mixtures**
· **Description:** Mixture of substances listed below with non-hazardous additions.
· **Dangerous Components:**

CAS: 1310-58-3	Potassium Hydroxide	25-50%
RTECS: TT 2102000	⬥ Skin Corr. 1A, H314; ⬥ Acute Tox. 4, H302	

US
(Continued)

Safety Data Sheet (SDS)
OSHA HazCom Standard 29 CFR 1910.1200(g) and GHS Rev 03.

Issue date 11/22/2019   *Reviewed on 11/22/2019*

Trade Name: Potassium Hydroxide, 40% Liquid

4 First-Aid Measures

- **Description of first aid measures**
- **General information:** Immediately remove any clothing soiled by the product.
- **After inhalation:**
  Supply fresh air. If required, provide artificial respiration. Consult doctor if symptoms persist.
  In case of unconsciousness place patient stably in the side position for transportation.
- **After skin contact:**
  Immediately wash with water and soap and rinse thoroughly.
  If skin irritation occurs, consult a doctor.
- **After eye contact:**
  Rinse opened eye for at least 15 minutes under running water.
  If easy to do so, remove contact lenses if worn.
  Seek immediate medical advice.
- **After swallowing:**
  Drink copious amounts of water and provide fresh air. Immediately call a doctor.
  Do not induce vomiting without medical advice.
  If vomiting does occur, repeat fluid administration
- **Information for doctor**
- **Most important symptoms and effects, both acute and delayed:**
  No further relevant information available.
- **Indication of any immediate medical attention and special treatment needed:**
  No further relevant information available.

5 Fire-Fighting Measures

- **Extinguishing media**
- **Suitable extinguishing agents:**
  CO_2, extinguishing powder or water spray. Fight larger fires with water spray or alcohol resistant foam.
- **For safety reasons unsuitable extinguishing agents:** No further relevant information is available.
- **Special hazards arising from the substance or mixture:** No further relevant information available.
- **Advice for firefighters**
- **Special protective equipment for firefighters:**
  As in any fire, wear self-contained breathing apparatus pressure-demand (NIOSH approved or equivalent) and full protective gear to prevent contact with skin and eyes.

6 Accidental Release Measures

- **Personal precautions, protective equipment and emergency procedures:**
  Wear protective equipment. Keep unprotected persons away.
  Avoid contact with skin, eyes and clothing.
  Ensure adequate ventilation.
- **Environmental precautions:**
  Dilute with plenty of water.
  Do not allow to enter sewers/surface or ground water.
- **Methods and material for containment and cleaning up:**
  Absorb with liquid-binding material (i.e. sand, diatomite, acid binders, universal binders, sawdust).

(Continued)
US

Safety Data Sheet (SDS)
OSHA HazCom Standard 29 CFR 1910.1200(g) and GHS Rev 03.

Issue date 11/22/2019   Reviewed on 11/22/2019

Trade Name: Potassium Hydroxide, 40% Liquid

Use neutralizing agent.
Dispose contaminated material as waste according to section 13.
Ensure adequate ventilation.
Dispose of the collected material according to regulations.
· **Reference to other sections:**
See Section 7 for information on safe handling.
See Section 8 for information on personal protection equipment.
See Section 13 for disposal information.

7 Handling and Storage

· **Handling**
· **Precautions for safe handling:**
Ensure good ventilation/exhaustion at the workplace.
Prevent formation of aerosols.
· **Information about protection against explosions and fires:** No special measures required.

· **Conditions for safe storage, including any incompatibilities**
Store away from strong acids, strong bases, strong oxidizing agents, strong reducing agents and amphoteric metals (Tin, Lead, Zinc, Aluminium).
· **Storage**
· **Requirements to be met by storerooms and receptacles:** Store in the original container.
· **Information about storage in one common storage facility:** Not required.
· **Further information about storage conditions:** Keep receptacle tightly sealed.
· **Specific end use(s):** No further relevant information available.

8 Exposure Controls/Personal Protection

· **Additional information about design of technical systems:** No further data; see section 7.

· **Control parameters:**
All ventilation should be designed in accordance with OSHA standard (29 CFR 1910.94). Use mechanical (general) ventilation for storage areas. Use appropriate ventilation as required to keep Exposure Limits in Air below TLV & PEL limits.
· **Components with occupational exposure limits:**
1310-58-3 Potassium Hydroxide
REL Ceiling limit value: 2 mg/m^3
TLV Ceiling limit value: 2 mg/m^3
· **Additional information:** The lists that were valid during the creation of this SDS were used as basis.

· **Exposure controls:**
· **Personal protective equipment**
· **General protective and hygienic measures:**
The usual precautionary measures for handling chemicals should be followed.
Keep away from foodstuffs, beverages and feed.
Immediately remove all soiled and contaminated clothing and wash before reuse.
Wash hands before breaks and at the end of work.
Avoid contact with the eyes and skin.

(Continued)

Safety Data Sheet (SDS)
OSHA HazCom Standard 29 CFR 1910.1200(g) and GHS Rev 03.

Issue date 11/22/2019 *Reviewed on 11/22/2019*

Trade Name: Potassium Hydroxide, 40% Liquid

· **Breathing equipment:**
In case of brief exposure or low pollution use respiratory filter device. In case of intensive or longer exposure, use respiratory protective device that is independent of circulating air.
· **Protection of hands:**

 Protective gloves

The glove material has to be impermeable and resistant to the product/ the substance/ the preparation. Due to missing tests no recommendation to the glove material can be given for the product/ the preparation/ the chemical mixture.
Select glove material based on penetration times, rates of diffusion and degradation.
· **Material of gloves:**
The selection of the suitable gloves does not only depend on the material, but also on further marks of quality and varies from manufacturer to manufacturer. As the product is a preparation of several substances, the resistance of the glove material cannot be calculated in advance and has therefore to be checked prior to the application.
· **Penetration time of glove material:**
The exact break-through time has to be determined and observed by the manufacturer of the protective gloves.
· **Eye protection:**

 Tightly sealed goggles

· **Body protection:**

 Protective work clothing

9 Physical and Chemical Properties

· **Information on basic physical and chemical properties**
· **General Information**
· **Appearance:**
  Form: Liquid
  Color: Colorless
· **Odor:** Odorless
· **Odor threshold:** Not determined.

· **pH-value @ 20 °C (68 °F):** >13

· **Change in condition**
  Melting point/Melting range: --
  Boiling point/Boiling range: 100 °C (212 °F)

· **Flash point:** None

(Continued)

Safety Data Sheet (SDS)
OSHA HazCom Standard 29 CFR 1910.1200(g) and GHS Rev 03.

Issue date 11/22/2019   *Reviewed on 11/22/2019*

Trade Name: Potassium Hydroxide, 40% Liquid

· Flammability (solid, gaseous):	Not applicable.
· Ignition temperature:	Not applicable
· Decomposition temperature:	Not determined.
· Auto igniting:	Product is not self-igniting.
· Danger of explosion:	Product does not present an explosion hazard.
· Explosion limits:	
Lower:	Not determined.
Upper:	Not determined.
· Vapor pressure @ 20 °C (68 °F):	23 hPa (17.3 mm Hg)
· Density @ 20 °C (68 °F):	1.416 g/cm³ (11.8165 lbs/gal)
· Relative density:	Not determined.
· Vapor density:	Not determined.
· Evaporation rate:	Not determined.
· Solubility in / Miscibility with:	
Water:	Fully miscible.
· Partition coefficient (n-octanol/water):	Not determined.
· Viscosity:	
Dynamic:	Not determined.
Kinematic:	Not determined.
· Solvent content:	
Organic solvents:	0.0 %
Water:	60 %
Solids content:	40 %
· Other information:	No further relevant information available.

10 Stability and Reactivity

· **Reactivity:** No further relevant information available.
· **Chemical stability:** Stable under normal conditions.
· **Thermal decomposition / conditions to be avoided:**
No decomposition if used according to specifications.
· **Possibility of hazardous reactions:** No dangerous reactions known.
· **Conditions to avoid:** No further relevant information available.
· **Incompatible materials:**
Strong acids, strong bases, strong oxidizing agents, strong reducing agents and amphoteric metals (Tin, Lead, Zinc, Aluminium).
· **Hazardous decomposition products:** No dangerous decomposition products known.

US
(Continued)

Safety Data Sheet (SDS)
OSHA HazCom Standard 29 CFR 1910.1200(g) and GHS Rev 03.

Issue date 11/22/2019　　　　　　　　　　　　　　　　　　　　　Reviewed on 11/22/2019

Trade Name: Potassium Hydroxide, 40% Liquid

11 Toxicological Information

· **Information on toxicological effects:**
· **Acute toxicity:**
· **LD/LC50 values that are relevant for classification:**

1310-58-3 Potassium Hydroxide
Oral　　　LD50　　　273 mg/kg (Rat)
Inhalative　LC50/96 hours　80 mg/l (Daphnia)

· **Primary irritant effect:**
· **On the skin:** Strong caustic effect on skin and mucous membranes.
· **On the eye:**
Corrosive effect.
Strong caustic effect.
· **Additional toxicological information:**
Swallowing will lead to a corrosive effect on mouth and throat and to the danger of perforation of esophagus and stomach.
The product shows the following dangers according to internally approved calculation methods for preparations:
Harmful
Corrosive
Irritant

· **Carcinogenic categories:**
· **IARC (International Agency for Research on Cancer):**
None of the ingredients are listed.

· **NTP (National Toxicology Program):**
None of the ingredients are listed.

· **OSHA-Ca (Occupational Safety & Health Administration):**
None of the ingredients are listed.

12 Ecological Information

· **Toxicity:**
· **Aquatic toxicity:** No further relevant information available.
· **Persistence and degradability:** No further relevant information available.
· **Behavior in environmental systems:**
· **Bioaccumulative potential:** No further relevant information available.
· **Mobility in soil:** No further relevant information available.
· **Additional ecological information:**
· **General notes:**
Do not allow undiluted product or product that has not been neutralized to reach ground water, water course or sewage system.
Must not reach bodies of water or drainage ditch undiluted or unneutralized.
Rinse off of bigger amounts into drains or the aquatic environment may lead to increased pH-values. A high pH-value harms aquatic organisms. In the dilution of the use-level the pH-value is considerably reduced, so that after the use of the product the aqueous waste, emptied into drains, is only low water-dangerous.

(Continued)

Safety Data Sheet (SDS)
OSHA HazCom Standard 29 CFR 1910.1200(g) and GHS Rev 03.

Issue date 11/22/2019   Reviewed on 11/22/2019

Trade Name: Potassium Hydroxide, 40% Liquid

- **Results of PBT and vPvB assessment:**
- **PBT:** Not applicable.
- **vPvB:** Not applicable.
- **Other adverse effects:** No further relevant information available.

13 Disposal Considerations

- **Waste treatment methods**
- **Recommendation:**
Must not be disposed of together with household garbage. Do not allow product to reach sewage system.
Observe all federal, state and local environmental regulations when disposing of this material.

- **Uncleaned packaging**
- **Recommendation:** Disposal must be made according to official regulations.
- **Recommended cleansing agent:** Water, if necessary with cleansing agents.

14 Transport Information

- **UN-Number:**
- **DOT, ADR/ADN, IMDG, IATA**   UN1814
- **UN proper shipping name:**
- **DOT**   Potassium hydroxide, solution
- **ADR/ADN**   UN1814 Potassium hydroxide, solution
- **IMDG, IATA**   POTASSIUM HYDROXIDE SOLUTION
- **Transport hazard class(es):**

- **DOT**

- **Class:**   8 Corrosive substances
- **Label:**   8

- **ADR/ADN**

- **Class:**   8 (C5) Corrosive substances
- **Label:**   8

(Continued)

Safety Data Sheet (SDS)
OSHA HazCom Standard 29 CFR 1910.1200(g) and GHS Rev 03.

Issue date 11/22/2019   *Reviewed on 11/22/2019*

Trade Name: Potassium Hydroxide, 40% Liquid

· IMDG, IATA

· Class:	8 Corrosive substances
· Label:	8
· Packing group:	
· DOT, ADR/ADN, IMDG, IATA	II
· Environmental hazards:	Not applicable.
· Special precautions for user:	Warning: Corrosive substances
· Danger code (Kemler):	80
· EMS Number:	F-A,S-B
· Segregation groups:	Alkalis
· Transport in bulk according to Annex II of MARPOL73/78 and the IBC Code:	Not applicable.
· Transport/Additional information:	
· DOT	
· Quantity limitations:	On passenger aircraft/rail: 1 L On cargo aircraft only: 30 L
· ADR/ADN	
· Excepted quantities (EQ):	Code: E2 Maximum net quantity per inner packaging: 30 ml Maximum net quantity per outer packaging: 500 ml
· IMDG	
· Limited quantities (LQ):	1L
· Excepted quantities (EQ):	Code: E2 Maximum net quantity per inner packaging: 30 ml Maximum net quantity per outer packaging: 500 ml
· UN "Model Regulation":	UN 1814 POTASSIUM HYDROXIDE, SOLUTION, 8, II

15 Regulatory Information

· *Safety, health and environmental regulations/legislation specific for the substance or mixture:*
· *SARA (Superfund Amendments and Reauthorization):*
· *Section 355 (extremely hazardous substances):*

None of the ingredients are listed.

· *Section 313 (Specific toxic chemical listings):*

None of the ingredients are listed.

· *TSCA (Toxic Substances Control Act):*

1310-58-3 Potassium Hydroxide
7732-18-5 Water, distilled water, deionized water

(Continued)

Safety Data Sheet (SDS)
OSHA HazCom Standard 29 CFR 1910.1200(g) and GHS Rev 03.

Issue date 11/22/2019   *Reviewed on 11/22/2019*

Trade Name: Potassium Hydroxide, 40% Liquid

· **California Proposition 65:**
· **Chemicals known to cause cancer:**
None of the ingredients are listed.

· **Chemicals known to cause reproductive toxicity for females:**
None of the ingredients are listed.

· **Chemicals known to cause reproductive toxicity for males:**
None of the ingredients are listed.

· **Chemicals known to cause developmental toxicity:**
None of the ingredients are listed.

· **Carcinogenic categories:**
· **EPA (Environmental Protection Agency):**
None of the ingredients are listed.

· **TLV (Threshold Limit Value established by ACGIH):**
None of the ingredients are listed.

· **NIOSH-Ca (National Institute for Occupational Safety and Health):**
None of the ingredients are listed.

· **GHS label elements**
The product is classified and labeled according to the Globally Harmonized System (GHS).
· **Hazard pictograms:**

· **Signal word:** Danger

· **Hazard-determining components of labeling:**
Potassium Hydroxide
· **Hazard statements:**
H290 May be corrosive to metals.
H302 Harmful if swallowed.
H314 Causes severe skin burns and eye damage.
· **Precautionary statements:**
P234	Keep only in original container.
P260	Do not breathe dusts or mists.
P264	Wash thoroughly after handling.
P270	Do not eat, drink or smoke when using this product.
P280	Wear protective gloves/protective clothing/eye protection/face protection.
P301+P312	If swallowed: Call a poison center/doctor if you feel unwell.
P301+P330+P331	If swallowed: Rinse mouth. Do NOT induce vomiting.
P303+P361+P353	If on skin (or hair): Take off immediately all contaminated clothing. Rinse skin with water/shower.
P304+P340	IF INHALED: Remove person to fresh air and keep comfortable for breathing.
P305+P351+P338	If in eyes: Rinse cautiously with water for several minutes. Remove contact lenses, if present and easy to do. Continue rinsing.

(Continued)

Safety Data Sheet (SDS)
OSHA HazCom Standard 29 CFR 1910.1200(g) and GHS Rev 03.

Issue date 11/22/2019   Reviewed on 11/22/2019

Trade Name: Potassium Hydroxide, 40% Liquid

P310	Immediately call a poison center/doctor.
P321	Specific treatment (see supplementary first aid instructions on this Safety Data Sheet).
P363	Wash contaminated clothing before reuse.
P390	Absorb spillage to prevent material damage.
P405	Store locked up.
P406	Store in corrosive resistant container with a resistant inner liner.
P501	Dispose of contents/container in accordance with local/regional/national/international regulations.

· **National regulations:**
None of the ingredients are listed.

· **Chemical safety assessment:** A Chemical Safety Assessment has not been carried out.

16 Other Information

The information and recommendations in this safety data sheet are, to the best of our knowledge, accurate as of the date of issue. Nothing herein shall be deemed to create warranty, expressed or implied, and shall not establish a legally valid contractual relationship. It is the responsibility of the user to determine applicability of this information and the suitability of the material or product for any particular purpose.

· **Date of last revision/ revision number:** 11/22/2019 / 1

· **Abbreviations and acronyms:**
ADR: The European Agreement concerning the International Carriage of Dangerous Goods by Road
ADN: The European Agreement concerning the International Carriage of Dangerous Goods by Inland Waterways
IMDG: International Maritime Code for Dangerous Goods
DOT: US Department of Transportation
IATA: International Air Transport Association
ACGIH: American Conference of Governmental Industrial Hygienists
EINECS: European Inventory of Existing Commercial Chemical Substances
ELINCS: European List of Notified Chemical Substances
CAS: Chemical Abstracts Service (division of the American Chemical Society)
NFPA: National Fire Protection Association (USA)
HMIS: Hazardous Materials Identification System (USA)
LC50: Lethal concentration, 50 percent
LD50: Lethal dose, 50 percent
PBT: Persistent, Bioaccumulative and Toxic
vPvB: very Persistent and very Bioaccumulative
NIOSH: National Institute for Occupational Safety and Health
OSHA: Occupational Safety & Health Administration
TLV: Threshold Limit Value
PEL: Permissible Exposure Limit
REL: Recommended Exposure Limit
Met. Corr.1: Corrosive to metals – Category 1
Acute Tox. 4: Acute toxicity – Category 4
Skin Corr. 1A: Skin corrosion/irritation – Category 1A
Eye Dam. 1: Serious eye damage/eye irritation – Category 1

· *** Data compared to the previous version altered.**
SDS created by MSDS Authoring Services   www.msdsauthoring.com   +1-877-204-9106

Appendix C: Factory Acceptance Testing Checklist

The checklist in this appendix is from ANSI/ISA-62381-2011 (IEC 62381 Mod), *Automation Systems in the Process Industry – Factory Acceptance Test (FAT), Site Acceptance Test (SAT), and Site Integration Test (SIT)*. It is reprinted with permission from ISA.

Annex A – Factory Acceptance Testing Checklist

The following typical tasks should be considered when preparing for Factory Acceptance Test (FAT) for a specific project:

1. Assemble all documentation applicable to the project prior to the FAT. Documentation may include:

    a. User requirements specification:

    - Automation system specification;
    - Automation system purchase order;
    - P&IDs;
    - Definition of plant operating units and equipment modules;
    - Color conventions for operator interface (display static and dynamic colors, alarm colors, running/off colors, other dynamic graphic block colors, etc.);
    - Alarm philosophy.

b. Functional design specification:

- Automation system functional design specification, including system architectural diagram(s), description of each major component and subsystem, block diagrams, etc.;

- System hardware specification, including all system and cabinet drawings showing grounding, power supply, wiring, interconnections, etc.;

- Vendor agreement(s) – e.g., purchase order, contracts;

- Instrument index and instrument specifications;

- Master system I/O list, including all tags connected to the system and all subsystems (e.g., SIS, PLC, unit controllers, analyzer subsystems, etc.);

- System functional requirements (see ANSI/ISA-5.06.01), including:

    – Operational interlock description/matrix;

    – Prose operating description (control narratives), including sequencing, abnormal modes of operation and failure contingencies;

    – Functional logic diagrams;

    – Description of complex control schemes, including a list of complex loops and tags with detailed descriptions, including methods of simulation with acceptance criteria;

    – Sketches of operator displays and symbols;

    – Assignment of plant units to operator workstations;

    – Alarm list, including priorities and settings;

    – Recipe parameters;

    – Formulae for automation system calculations;

    – Reporting requirements, including format;

    – System security and access requirements;

    – History collection definition, including history server capacity, spare capacity, removable media or backup method.

c. Information to vendor: Any updates, clarifications, or additions to the requirements and information provided to the vendor during the execution of the order shall be included in this documentation.

d. Vendor documentation: Documentation developed by the vendor shall also be gathered and become part of the FAT reference documentation. This may include, but is not limited to, the following:

- Manuals (operating, installation, etc.), system data sheets, certificates;
- Layout drawings and bill of materials;
- Assembly and wiring drawings;
- Installation drawings and instructions;
- System architecture diagram, including system device addresses;
- Processor loading calculations, network loading reports, and server loading reports;
- Power distribution panel – system power loading calculations, UPS load requirement, MCB ratings, test reports etc.;
- Description of interfaces to other systems/components not part of this purchase order;
- Print outs of graphics;
- Final Input/Output list;
- Configuration (program) print out;
- Vendor test reports;
- Software licenses with release/version information and tag capacity;
- Revision of configuration/application program at start of FAT;
- Any checklist used during the pre-FAT and FAT and all test procedure reports;
- List all protocols used for external communication;
- Documentation of qualifications/certifications of vendor personnel performing specialized work (e.g., "code welding").

2. Develop a written test plan and specification. The following provides typical tests for various aspects of a system that should be included in the test plan. A test plan should be developed for each system that tests and verifies the system's compliance with the project specifications and purchase order, including the following:

a. Documentation

- All necessary documentation should be available and current.

b. Hardware/software inventory:

- Hardware: architecture, bill of materials, dimensions, painting, etc.;
- Software: licenses/versions including firmware;
- Spares, consumables, and tools;
- Hardware and software configuration parameters.

c. Mechanical inspection:

- Mounting of components and modules;
- Spare capacity;
- Maintainability of cabinet fans, construction of cabinets;
- Cable entry, support bars and accessories (cable clamps, glands, etc.);
- Earthing, equipotential bonding;
- Screwed connections, terminal connections;
- Labeling, tagging of components;
- Electric shock protection, warning labels.
- Mechanical protection of items in order to prohibit inadvertent operation.

d. Wiring and termination inspection:

- Wiring and cabling;
- Fusing, circuit-breakers;
- Tagging, labeling of cables, wires, etc.;
- Segregation and identification of different wiring (e.g., AC, DC, intrinsically safe, etc.);
- Wire crimp inspection (random) and manual wire pull test (random);
- Cable duct loading;
- Power supplies and power distribution.

e. Start-up test and general system functions:
   - Cold start up;
   - On-line change to configuration/program;
   - Controller cycle time;
   - Display call-up time;
   - Value update time;
   - System load (memory capacity, storage capacity, etc.);
   - Time synchronization check;
   - Check of user log-on security (if any) for operation and control;
   - System security levels and user logons;
   - Demonstrate battery backup operation.

f. System alarm test:
   - Alarm processing strategy and acknowledgement;
   - Power-supply failure, UPS monitoring and diagnostics;
   - Fuse, breaker monitoring;
   - Cooling fans, high enclosure temperature alarm;
   - Communication, network monitoring;
   - Short circuit, wire break, out of range, earth fault;
   - Watchdog, if any;
   - Priority levels of alarms;
   - System diagnostics.

g. Hardware redundancy and diagnostics:
   - Redundant operation and monitoring of controllers;
   - Redundant operation and monitoring of communication and networks, including network switches;
   - Redundant operation and monitoring of power supplies;
   - Redundant operation and monitoring of operator stations;

- Redundant operation and monitoring of I/O, if any;
- Redundant operation of third party systems;
- Redundant operation and monitoring of all other devices not mentioned before.

h. Operator interface: Prior to any loop-oriented or I/O-oriented test, the static parts of the HMI displays (if applicable) shall be verified. The following display functionality (static and dynamic) shall be verified, using the applicable P&ID or display requirements sketch:

- Symbols for vessels, process lines, valves, transmitters, motors, pumps, etc.;
- Colors for static items, for example, hand valves, process lines, etc.;
- Process flow direction and path, i.e., process line arrows;
- Hierarchies and all linking of displays;
- The designed dynamic changes of colors, sub-pictures and data entry points;
- Density of information (static and dynamic) contained on a display;
- Colors of background and color changes;
- Static text and dynamic changes;
- Organization (jumps, transitions, sub-pictures).

i. Engineering workstation functionality shall be tested as follows:

- Generating draft graphics;
- Loop configuration/changes;
- Generating reports & logs;
- Using text editor;
- Using database builder;
- Testing historization module.

j. Test I/O to the operator interface display: Each I/O point in the system shall be tested as follows:

- The faceplate for the I/O point shall be checked to confirm required appearance and functionality:

- Functionality, service text, range, units, state text, etc.;
- Link to proper physical I/O point/address;
- Correct linking of split range control schemes;
- Related group display and process graphic display;
- Related trends.

- It shall be verified that the tag target on the graphic(s) is(are) in the correct location and that the color changes for dynamic targets, for example, valves, motors, bar graphs, etc. are correct.

- Check of alarm assignments – Type, value, sorting criteria (priority, plant area, etc.).

k. All I/O shall be tested for proper operation to the fullest extent possible, including verification of individual I/O ranges and engineering units.

The most complete I/O testing method is:

- Forcing or manipulation of I/O by means of simulation devices connected at field terminals (thus including in the test marshalling terminals, interface devices (e.g., barriers), cross-wiring, system cabling, and I/O modules).

Alternate I/O testing methods, not as encompassing as the above test, are listed below to be considered only when the above complete test is not possible for some specific reason.

- Forcing or manipulation of I/O by means of simulation devices connected to the system I/O modules to verify correct performance of that I/O and all associated displays, alarms, etc.;

- Forcing or manipulation of I/O by means of software simulation of the I/O modules;

- Forcing or manipulation of I/O by means of software simulation at the processor level.

l. Bus interfaces shall have a generic test conducted for each specified type of field or external device bus interface which is compliant to the relevant standard. This test should cover the interoperability of the automation system and the device.

- One bus segment should be built up and tested with all associated devices linked to it. Selection of the segment to be tested should be mutually agreed upon and should generally attempt to prove function of worst installed case;

- In the case of BPCS functionality, all concerned segments of the BPCS should be tested or simulated;

- Signals related to segments not built up should be simulated;

- All relevant documents, data sheets, figures (load, cycle time, architecture) should be reviewed for all segments.

m. Any portions of the system hardware not fully tested during the FAT (e.g., marshalling terminals) should be fully tested as part of the SAT and any issues resolved at that time.

n. Check of complex functionality and interlocks: The test of complex functionality and interlocks should be carried out after the I/O tests. The tests to be performed should be as outlined in the FAT document and include:

- Test of complex functions;

- Operational interlocks within the BPCS;

- Project reporting requirements.

o. Test of communication links to subsystems: Whenever possible, every connection to an external system should be tested. If the actual connection to the external system cannot be made during the FAT, the test should be performed by means of a subsystem simulation device. Actual signal type/level, signal architecture, for example redundancy, and connection should be provided as far as practicable. The manner of testing should be defined for each subsystem individually to demonstrate compliance with the project requirements. The simulation of signals and checking of all functions related to automation subsystems should be carried out in accordance with the project FAT test specification. In addition to the application-related test for correct functionality, all other applicable system features should be checked, such as:

- Effect of link failure

- Recovery from failure

- Redundancy

- Alternative modes of operation

p. Check of other system functions: In addition to the application-related test, other system should be tested, including:

- Recovery from failure;
- Vendor guaranteed system performance (refresh rate, etc.);
- System functionality configured by the vendor but not defined/required by the owner;
- All other functions not otherwise specifically listed here.

3. Develop a test schedule including, but not limited to, the following typical activities:

Item	Description
1	Initial meeting (FAT document and plan review, scope of tests, schedule, etc.)
2	Vendor documentation (including in-house test reports) check
3	Hardware and Software inventory check
4	Mechanical inspection
5	Wiring and termination inspection
6	Start-up test and general system functions
7	System alarm test
8	Hardware redundancy and diagnostics
9	Operator interface
10	IO tests from field wiring terminals to I/O terminals, to controller and including HMI.
11	Complex functionality and operation modes (for example, batch, sequence control)
12	Test of Communication Links to Subsystems and third party interfaces.
13	FAT rework, punch list development
14	FAT close-out meeting

Appendix D: Answers to Review Questions

Chapter 1

1.1. With regard to wiring, what types of high-quality work must a CST perform?

Whether the CST performs this work or inspects it, good housekeeping, proper termination of wiring, and good practices in running wire, installing wire tags properly, and maintaining documentation are examples of high-quality work for which a CST is responsible.

1.2. Give examples of situations in which a CST may serve as an assistant during plant design.

A CST may serve as an assistant (1) during process safety reviews, (2) when helping design control schemes to make product in the new plant and to avoid unsafe occurrences or accidents, and (3) when interfacing with staff from engineering firms who are not as knowledgeable about the plant as the CST is.

1.3. What would you consider to be essential tasks for a CST during start-up?

Loop checking, calibration, and process automation system (PAS) configuration (to name a few).

1.4. Why can a CST function as an effective safety and quality inspector?

The CST has experience and expertise and should have good communication skills. It is beneficial for a company to use people "in house." The CST might be asked to attend a process hazard analysis (PHA).

1.5. How do vendors get involved during the start-up?

Vendors normally help start up equipment made by their company that has been purchased for the new plant. Their specialists are typically very knowledgeable about this equipment, and they can provide the equipment's operating manuals, cut sheets, and other documentation and discuss them with you.

1.6. What are some purposes of a job description?

A job description provides an employer with a clear means of explaining the position requirements on a regular basis. It states what a CST does and is responsible for including the necessary skills, duties, responsibilities, training, and education required. Job descriptions can be used when hiring to review the position with a potential employee, to orient new employees to their positions, and to evaluate current employees' job performance within the company.

1.7. Who might report to a lead CST, and what are a lead CST's responsibilities?

Other technicians, such as instrument technicians, electricians, junior CSTs, or apprentices, may report to a lead CST. The lead CST coordinates their activities and paperwork, and attends meetings to discuss work. The lead CST also instructs these people while still performing the tasks covered by the CST's job description.

1.8. What does a CST do when serving as a liaison between vendors and/or contractors and the other members of the start-up team or plant organization?

When serving as a liaison, the CST may function as an escort, as someone who helps procure materials (safety and mechanical) for visitors, or as the person who relates information to the project manager (and vice versa).

1.9. What level of education and training do you think a CST should have in order to be involved in a start-up?

Examples of necessary education and training activities for CSTs involved in start-ups include trade school, additional off-site training on a PAS, and apprenticeships. ISA certification is available too.

1.10. What is meant by *baseline work*?

Baseline work *refers to the basic level of duties and responsibilities that the CST will be involved in on a daily basis during start-up.*

1.11. In which subjects do you think a CST could teach, train, or instruct people?

The CST could train others in subjects such as safety on the job, calibration and loop checking procedures, and standard procedures for performing a CST's tasks.

1.12. Why is it important for a CST to communicate frequently with vendor representatives and specialists?

Communicating with vendor representatives and specialists gives the CST the opportunity to gain expert knowledge in how a system works and how to troubleshoot it from the most knowledgeable person associated with the system.

Chapter 2

2.1 A calciner with a local burner management system (BMS) has a local panel controlled by a programmable logic controller (PLC). The flame does not stay lit. Someone is suggesting that the purge time is too long and that the pilot is being blown out. He suggests testing by "jumpering out" an interlock to prove this. What does he mean by this? Why is this wrong? How can you fix this problem? What documents might you use to work on this job?

Jumpering out a circuit refers to bypassing the circuit with a piece of wire. This is unsafe—the circuit is there for a reason. Sometimes this jumper is forgotten (left undocumented) and remains intact until a problem occurs and someone else has to work on it. If this part of the electrical scheme is disabled, then the unit's design has changed and safety may be compromised. This also changes the standard operating procedure (SOP) for the unit. You may be able to step through the PLC program manually to see what the problem is (as this is slower, it allows you to see what is happening step by step) or call the vendor and have him or her come in if necessary.

There may be an error in the program, and someone will need to reprogram the controller, run the unit on low fire, and see if the flame remains intact (do this in manual if safety is not compromised). Check the loop sheets, electrical, and vendor drawings.

2.2 You have been told you must attend a process hazard analysis (PHA) meeting about a change occurring in Area 3 of the plant. What should you do to prepare for the meeting? What might you bring to the meeting? Why were you invited to this meeting?

Look at the process flow diagram (PFD) and piping and instrumentation drawing (P&ID) before the meeting to understand the process. Visit the plant area, talk to operators about their operation, and review the SOP. Bring a notepad, a pen or pencil, and the drawings you have been working with to the meeting. Bring electrical drawings as necessary. You are considered an expert in your area (control systems and instrumentation). You will be a great asset in the meeting as questions come up that you can help answer. As a result of your professional background, you will also raise questions that others may not have considered.

2.3   It is 3:30 p.m. on Friday. The plant operations manager has decided that he wants a 150 hp (111,855 W) blower replaced with a 300 hp (223,710 W) blower because not enough airflow is being provided to the fluid bed dryer. He thinks the job should not take more than about 4 hours, and he wants production to be back online by 11 p.m. What are some of the factors that might compromise the safety of this job? What should be done to ensure that maintenance personnel can work on the blowers safely? What might you as a control CST be involved with in connection with this task?

Possible safety factors include pressure to do the job quickly, pressure from management, lateness in the day and week (and consequently, tiredness), and the possible lack of adequate help to accomplish the job. To ensure the safety of maintenance personnel, use proper lockout/tagout procedures, including permits; make sure MOC has been initiated; and get necessary documentation to do the job. The CST may need to loop check the new blower if any wiring has changed; if there are transmitters associated with this process, their ranges and elements may need to be changed (e.g., flow must be increased).

2.4   What government organization created the *Process Safety Management of Highly Hazardous Chemicals* standard?

The US Occupational Safety and Health Administration (OSHA)

2.5   What is the objective of process safety management (PSM)?

The objective of PSM is to prevent the unwanted release of hazardous chemicals (especially into locations that could expose employees and others to serious hazards).

2.6   Name two engineering societies that provide technical reports that help maintain good engineering practice.

The American Institute of Chemical Engineers (AIChE) and the International Society of Automation (ISA)

2.7   What document gives information for determining which chemicals are hazardous and how to deal with them safely or eliminate them completely?

A Safety Data Sheet (SDS) is provided by the manufacturer of the chemical. This document shows chemical properties, hazards, and the required personal protective equipment (PPE).

2.8   Name some drawings that are important during a PHA and pre-startup safety review (PSSR).

Important drawings include PFDs, P&IDs, loop sheets, and miscellaneous wiring diagrams.

2.9 Is following Management of Change (MOC) procedures required when replacing a pump with another one of the same type?

No, following MOC procedures is not required for a "replacement in kind."

2.10 Why is it important to consider materials of construction when working on a job or replacing parts?

Materials that are incompatible with process chemicals could fail, resulting in equipment damage and environmental and safety problems (e.g., chemical leakage).

2.11 What is another term for *nonroutine work*?

Hot work

2.12 What is the sequence of events necessary for the proper lockout of a pipeline attached to a pump possibly filled with a (liquid) chemical?

Shut down the operation, de-energize the pump power, install a lock(s) on the pump motor starter(s), test for power before starting work, block and bleed lines/process, and drain lines and install blanks/blinds if necessary.

2.13 Name two important documents used during lockout.

Permit and tag

2.14 Identify some documents that would need to be changed as a result of a process change.

The PFD, P&ID, and SOP would need to be changed. There could be several others including stores inventory, paperwork, and electrical drawings.

2.15 Name some teams (including volunteer organizations) a CST may be involved with.

Teams a CST may be involved with include emergency response teams (ERTs), safety PHA and PSSR review teams.

2.16 Define a *near miss*.

A near miss *is a potential for injury to personnel or damage to equipment*

2.17 Name some types of safety training that plants offer.

Regulatory, plant-specific, home and personal, self-study, hands-on, and computer-based training may be offered.

2.18 What training is mandatory? Name some of the topics covered by such training.

OSHA regulatory training is mandatory. This training covers topics such as PPE, emergency preparedness, and MOC.

2.19 Why can start-ups be particularly hazardous?

There are many people in the plant at the same time; people are nervous or in a hurry to get things done; people may be suffering from a lack of sleep; there may be management pressures; and so on.

2.20 Name some safety documents.

SDSs and safety audits are typical safety documents.

2.21 Name five pieces of PPE.

Examples of PPE include safety glasses, hard hats, steel-toe shoes, ear plugs, Nomex, and face shields and respirators.

Chapter 3

3.1 What are the differences between a piping and instrumentation drawing (P&ID) and an installation detail?

A P&ID is the primary schematic drawing used for laying out a process control installation. Major equipment, piping, and instrumentation are shown along with all connections—pneumatic, electrical, hydraulic, electromagnetic, and logical (software). Other data describing materials of construction, pump size, and head and impeller size may be shown in blocks at the bottom of the P&ID.

An installation detail shows where and how equipment should be installed in the plant so it works effectively. It is often to-scale and generic for the type of equipment being installed so it applies to different types of installation, not just to this specific plant.

3.2 In the past, what material was used for P&IDs and other engineering drawings?

Vellum

3.3 What is the advantage of scanning documents?

Scanning means you do not have to redraw anything.

3.4 Which document is used to track project activities and status and to coordinate between various activities?

The Gantt or "critical path" chart

3.5   What does SOP mean, and why are SOPs important?

A standard operating procedure (SOP) is important because the existence of such a document and the procedures resulting from its use can prevent errors from occurring. SOPs can also ensure more consistent practices and more consistent use of equipment, resulting in a consistent product and higher quality control.

3.6   Name some of the standards used when drawing documents.

Instrument and equipment symbology are two of the standards used—valve symbols are just one example. Some standards include information on how the line designations are made, showing materials of construction. Many documentation standards are developed and published by ISA and may be supplemented by standards at individual plants. Standards that are established and used worldwide are known as industry practices (IP).

3.7   What are the disadvantages of scanning documents?

In many systems, a scanned document becomes a "flat file." One does not have access to each word and cannot change the text. More recent scanning technologies, including optical character recognition (OCR), may address this problem.

3.8   What are the "layers" used for on computer-based P&IDs?

Different information may reside on different layers of the P&ID. For example, temporary changes or in-progress work that can be easily deleted may reside on a layer separate from the permanent plant equipment layer. Additionally, one group of people or one technical discipline may want only one type of information and may not need to print out or look at all the layers.

3.9   What does EDMS stand for?

Electronic document management system

3.10 Why is equipment manufacturer information important? What can these documents tell us?

An equipment manufacturer should know the equipment it builds—both the specifications and the tolerances of it—very well. The manufacturer can therefore make recommendations to maintain the equipment for a longer life and better operation. Some of the information equipment manufacturers provide includes installation instructions, replacement and preventive maintenance procedures, and start-up order (e.g., slowly heat up/cure the brick properly). Manufacturers can also provide troubleshooting and technical documents.

3.11 What are "bubbles" used for on engineering drawings?

"Bubbles" or "clouds" are elements added to a document to show changes from previous documents. They also denote work that must be done or is incomplete or uncertain (on HOLD). Bubbles are removed as the plant is updated and the changes are integrated.

3.12 Name some locations where documents are normally found.

Documents are typically found in control rooms; engineering offices; the Maintenance, Planning, and Purchasing departments; and technical libraries.

3.13 What is an energy balance?

An energy or heat balance is a mathematical formula that describes the process energy inputs and outputs that must be equal. The First Law of Thermodynamics states, "Energy is neither created nor destroyed," and is shown mathematically as follows:

$$\Sigma \text{ Inputs} = \Sigma \text{ Outputs or } \Sigma \text{ Inputs} - \Sigma \text{ Outputs} = 0,$$
where Σ is the symbol for "the sum of"

This reads, the sum of all inputs equals the sum of all outputs.

3.14 What are the differences between pneumatic, electrical, hydraulic, electromagnetic, and logical connections? How are these shown on P&IDs?

Referring to the following table:

1. *Pneumatic devices are air driven. On P&IDs, pneumatic signals are designated by a line with two hatch marks.*

2. *Electrical connections often use 4–20 mA analog and 110 V digital electrical signals. Electronic and logical connections are controlled digitally by computers. Electrical signals are dashed lines.*

3. *Hydraulic connections are oil or fluid driven and are indicated by a line hatched with L-shaped marks.*

4. *Electromagnetic signals are indicated by a line hatched with squiggly marks.*

5. *Logical (software) connections use dashes and circles.*

Number	Symbol	Application
1	—//——//—	• Pneumatic signal, continuously variable or binary.
2	— — — — —	• Electronic or electrical continuously variable or binary signal. • Functional diagram binary signal.
3	—L——L—	• Hydraulic signal.

4	⌒⌒	• Guided electromagnetic signal. • Guided sonic signal. • Fiber optic cable.
5	—o———o—	• Communication link and system bus, between devices and functions of a shared display, shared control system. • DCS, PLC, or PC communication link and system bus.

3.15   What types of controllers use Ladder Logic?

Programmable logic controllers (PLCs) often use Ladder Logic.

3.16   What is a Gantt chart? How does the CST contribute to the Gantt chart?

The project Gantt chart, often referred to as the project schedule, *is used during project design and up to and through start-up to track project activities and status as well as to coordinate various activities. It is usually a graphical means of depicting the parts of the total project, the length of time each part will take to complete, and how the parts interact with each other during the project and the subsequent commissioning and start-up of the plant. The Gantt chart is broken into the parts of the project, such as groundwork, construction, and start-up. As a CST, you contribute to the Gantt chart by getting your own work done, communicating to management that work is completed, or explaining that there are problems that may cause delays. The Gantt chart is updated based on your input. Gantt charts are also used in other situations, such as plant turnarounds, and for project planning.*

Chapter 4

4.1   A CST becomes aware that several valves from a manufacturer are not working properly. What should the CST do to prepare for the next project review meeting?

The CST should document the problem and bring the information to the start-up team's project review meeting. During these discussions, the team will determine what should be done, and the project schedule may have to be modified. Additional resources may be required to solve the problem or to work around it. If the meeting is not going to be held for a while, then the CST should approach the plant project person who is in charge of the valves or who purchased them and explain the situation. This person may need to use the CST's documentation or have the CST meet with the vendor to discuss the issue. Modifications to the valves or replacements may be discussed and parts ordered or replacements made. The CST will probably be involved in the entire process.

4.2 Why is it important for the CST to communicate work status to the start-up team at the project review meetings?

The CST communicates that the job is complete or that help is required to get the job done. The CST may be present at the project review meeting and will be required to discuss relevant parts of the job to provide status updates to the project team.

4.3 What crafts would normally be included in the Maintenance department?

The instrument, electrical, and mechanical (pipe fitters, machinists, etc.) crafts would be included.

4.4 Why does the CST interact with so many departments, groups, and individuals during a start-up?

All disciplines are involved in the start-up. The CST is an integral member of the team who contributes greatly to a successful start-up. The CST must communicate with all disciplines because each has a special role to play in starting the plant up.

4.5 Why is it important for the CST to attend project start-up meetings?

The CST must communicate his or her progress and that of his or her department toward project completion as well as any problems that have been encountered.

4.6 Why is it important to ask for help?

No one knows everything. Each person involved in the project is a specialist in his or her field. Asking for help will save time.

4.7 What is a *contractor liaison*?

A contractor liaison *is the staff person who communicates between the company at which the start-up takes place and the contractor doing the work. This person may help get work done or get parts or hot work permits for the contractor.*

4.8 Which government agencies does the Environmental department come in contact with?

The Environmental department comes into contact with federal and local EPA and departments of air and water quality.

4.9 Why is the Purchasing department an integral part of a start-up?

The Purchasing department is responsible for the paperwork used to order materials, parts, and services. They are involved in getting the bills paid. They have leverage (payment) and can withhold payment if materials are defective or services ineffective. In the bid process, they get competitive pricing for services and products.

4.10 Which discipline is the CST most likely to work with for basic process control system (BPCS)/distributed control system (DCS) work?

The control systems engineer (CSE)

4.11 What type of engineer is normally involved in mass and energy balances?

The process engineer

4.12 Why might the Human Resources (HR) department become involved in a start-up?

Contract issues such as manpower, overtime, payment, and safety may necessitate the involvement of the HR department.

4.13 What types of contractors might be involved in a start-up?

Electrical and mechanical contractors, construction companies, and control system integrators may be involved in the start-up.

4.14 What types of vendors would be involved in a start-up?

Vendors of pieces of major equipment (dryers, filters, tanks, etc.) and control systems, as well as raw material suppliers would be involved.

4.15 What is meant by the *chain of command*? Why is it important?

The chain of command *represents the order in which information and decisions are conveyed. It is important because everyone who must be informed and those who have the power to make decisions are involved.*

4.16 Who can a CST ask for help?

You should feel free to ask almost anyone for assistance. Engineers are happy to help and will need help from you, too. You may call vendors for technical advice and ask other plants within the corporation for assistance, particularly if they have a similar process or use the same types of equipment and instrumentation. Asking for help educates you and helps you find answers quickly.

4.17 What is the difference between *project engineers* and *process engineers*?

A project engineer *is normally involved in the construction and design of major equipment including pumps. This engineer is also responsible for the costs associated with the project and for getting all disciplines involved in its completion.*

The process engineer is normally involved in plant design from a capacity and quality standpoint. He or she is involved in handling mass and energy balances, sampling, and throughput.

Chapter 5

5.1   Why is it important to follow good Management of Change (MOC) procedures during start-up activities?

It is important to follow MOC procedures to ensure that everyone knows what is happening (i.e., to have good communication), and to comply with Occupation Safety and Health Administration (OSHA) process safety management (PSM) to ensure safety and efficiency and maintain records of what changed.

More activities converge as depicted on the "critical path" as the plant is nearing completion. These activities influence each other to a great degree, and many people must work in the same area. Safety can be compromised. The time for production demands is getting closer, and the project may be late getting started so people are getting impatient, nervous, and tired. It is important to keep everyone informed to maintain safety and efficiency. This is accomplished by maintaining good lines of communication, holding meetings more frequently, and issuing more memoranda.

5.2   Describe which documents would need to be updated if a new pipeline with a valve and a pump were added to a process.

A new Management-of-Change (MOC) form would be required to indicate the change and begin the change process. Updates would most likely be required to the process flow diagram (PFD), piping and instrumentation drawing (P&ID), and general arrangement (GA) drawing. Additions would probably be needed for electrical drawings, possibly Ladder Logic and other software programs. A new loop sheet, instrument specification sheet, and calibration sheet would also probably be necessary.

5.3   Describe a change that involves adding a new chemical that would initiate the MOC process.

An example is adding a new line as well as associated valves and pumps. An MOC form should be filled out and appropriate signatures obtained. A process hazard analysis (PHA) may be conducted (see Chapter 2). Additional training may be necessary for operators and maintenance personnel (including the CST). A scenario similar to a start-up (loop checking and calibration as well as hydrotesting) will probably occur.

5.4   What is the red pencil used for when marking up a piping and instrumentation drawing (P&ID)?

The red pencil is used to add items to a drawing. For example, if a line is to be modified on a drawing, you would "scratch it out" with the green pencil then redraw it with the red pencil. Red = add, green = remove.

5.5 Why is it important for the CST to understand MOC procedures?

To protect his or her safety, the safety of others, and the safety of company assets (equipment).

5.6 Who is responsible for generating the spec sheets?

The plant or outside engineering firm that is responsible for or was commissioned to create most of the drawings for the start-up and has done the engineering design for the new plant is normally responsible for generating the spec sheets. Sometimes the instrument vendors provide these too.

5.7 What type of information is on an ISA standard spec sheet?

The following information is on an ISA standard spec sheet: the type of instrument, manufacturer, size, device nomenclature, ranges for calibration, and use. See ISA-20-1981, Specification Forms for Process Measurement and Control Instruments, Primary Elements, and Control Valves.

5.8 What is *redlining*?

Redlining *is the term used to describe the process of correcting a document (by hand) by drawing information such as the current installation, configuration, calibration range, and location on it. It is called redlining because red is normally the color used to add information to engineering drawings. Green is used to delete information.*

5.9 What are *as-builts*?

As-builts *are drawings that reflect how the plant is in reality (currently) after changes have occurred to the original design. The original drawings must be updated to show the additions, deletions, and changes.*

5.10 Is there an ISA standard document for loop checking?

There is no ISA standard document for loop checking. Some plants use ISA standard calibration sheets during loop checking; others use these in conjunction with P&IDs and loop check log sheets (a list of tags to be checked) all contained within the loop folder.

5.11 What does the Gantt chart have to do with calibration?

The Gantt chart is important for keeping the calibration efforts on schedule to meet the start-up deadlines. This is because calibration impacts installation and loop checking, which are both necessary before start-up.

5.12 Why is good record keeping, including MOC, important?

Good record keeping promotes good communication and efficiency and provides plant documentation for future changes. It also helps prevent accidents, or if an accident occurs, it ensures that information will be available to help determine what happened and prevent it from happening again.

Chapter 6

6.1 What are some productive things a CST can do during slow (idle) times?

Talk with the board operator about loops that may not be controlling as expected, tune loops where required, and learn more about the plant.

6.2 Why is it important to have people cross-trained in preparation for a start-up?

Cross-training ensures staffing flexibility for coverage when someone is not available due to the legal limitations on the number of hours a person can work, illness, jury duty, illness or death in their family, or other reasons. Even when a trained person is on hand, sometimes "two heads are better than one" in a troubleshooting situation.

6.3 Why might the number of CST personnel involved in a start-up differ from plant to plant?

Plant size and process type affect the number of instrument and electrical (I&E) and CST personnel involved in the start-up. If the facility is completely new to the site, more people will be involved than if it is an addition to an existing facility. Some CSTs may be needed in other parts of the existing facility before they are needed in the new part, so they may not become involved in the start-up.

6.4 What is a *lead technician*, and what are this technician's responsibilities?

The lead technician is a more experienced instrument tech or is possibly a CST (which may be a requirement). He/she functions as a leader by monitoring other technicians' work and coordinating their efforts; is responsible for reports on plant and start-up status; assigns and follows up on the work that his or her team is performing; performs installation; and does loop checking, configuration, and troubleshooting.

6.5 What jobs might a CST be involved in?

A CST might be involved in calibration; equipment and device installation and system connection; loop checking; configuration; and troubleshooting.

6.6 Why does the quantity of work required of the CST vary throughout the start-up?

The quantity of work varies during start-up because events can take longer than anticipated. Even after the plant has been started up, CSTs can still be required because plant management may want to staff the plant around the clock with "experts" in the event that problems occur.

6.7 What things must be completed before a vendor representative arrives at the plant?

Electrical, pneumatic, and controls installation should be completed. People should be trained. Loop checks should be completed. Grounding per the manufacturer's instructions should be complete.

6.8 What are some skills or job functions that a CST may receive cross-training in?

The CST may receive cross-training in electrical wiring and installation, as well as computer, distributed control system (DCS), and programmable logic controller (PLC) configuration, installation, and replacement.

6.9 What safety factors should be considered in connection with the process automation system (PAS) when preparing for start-up?

The DCS and PLC equipment should be powered, grounded, and configured (programmed). Field equipment must be installed, powered, calibrated, and loop checked. Paperwork should be completed and communication should be coordinated with all involved to maintain a safe work environment.

6.10 How are the CST's assignments and responsibilities determined?

The CST's assignments and responsibilities are determined by the following: the CST's immediate supervisor, project review meetings, the project Gantt chart or schedule, and/or the plant management's decisions. These factors are normally dictated by marketing requirements.

Chapter 7

7.1 Provide examples of how the Operations department can give you input toward solving a problem.

Operators can explain the problem they are experiencing and give examples of what they are seeing or how the instrument used to perform. They may be able to offer a solution to the problem. They can assist in diagnosing the problem by helping the CST perform loop checks and by helping him or her maintain a safe work environment. They can help prove that the solution for the problem works. The operator can explain what impact the problem has on his or her job and on the operation of the plant.

7.2 Name some items a quality assurance (QA)/quality control (QC) inspector might find that should be added to a punch list?

An inspector would note any of the following:

- *Poor housekeeping*
- *Unterminated wiring*
- *Unlabeled enclosures*
- *Conduits that are not closed*
- *Missing manhole covers*
- *Unsecured or inaccessible instrumentation*

7.3 How do calibration and loop checking influence rework during commissioning and start-up, and how do they affect the timing of commissioning and start-up?

For calibration and loop checking, rework can be avoided if detailed procedures are set up and adhered to and paperwork is properly filled out. It is also important that qualified personnel do this work and ensure that the work is completed before the next phase of the start-up is planned. The entire project, including start-up, can be delayed by rework and sloppy work efforts.

7.4 Why do vendors charge for calibration at the factory?

Vendors charge for calibration at the factory because there are expenses associated with this type of work: labor, materials for calibration, and the necessary work areas.

7.5 Why is it important to tag instruments as "calibrated" after they have been calibrated?

It is important to tag instruments as "calibrated" so they are not confused with uncalibrated equipment. Tagging makes it clear that they are ready to be installed or used. It is probably part of a standard operating procedure (SOP). The project and start-up could be delayed if this type of confusion is introduced.

7.6 How does calibration methodology affect ISO 9000:2015 certification?

Appropriate standards and the proper maintenance of paperwork, including the use of trained personnel to perform the calibrations, are important for attaining and maintaining ISO 9000 certification.

7.7 What standard document is referred to for engineering design, calibration, and loop and function checks?

ISA specification (spec) sheet

7.8 Where might calibration and loop and function check documents be available?

These documents might be available in the control room, instrument shop, Engineering department, electronic document management system (EDMS), or PC network server(s).

7.9 Why is it important that qualified personnel be involved in calibration and loop and function checking efforts?

Some examples are to ensure that a quality job is done, to complete the job efficiently, to attain ISO 9000 certification, and to ensure safety and correct operation.

7.10 What are some examples of reasons instruments may need to be recalibrated?

Changes in process flows or parameters, errors in design, errors in calibration, or poor installation are a few examples.

7.11 How do you think calibration documentation affects Occupational Safety and Health Administration (OSHA) process safety management (PSM) compliance?

Paperwork associated with calibration is part of OSHA PSM points #8 Mechanical Integrity and #10 Managing Change. This paperwork must be complete to comply with the OSHA PSM standard and will be checked during audits.

7.12 Why is it important to have good communications skills when working on a problem during start-up?

The information and assistance you will need to solve a start-up problem will come from the operator and/or foreman as well as other instrument and electrical (I&E) technicians in the area. Therefore, it is vital to have the ability to ask questions and listen to the answers. It is also important for you to provide the much needed information you have. This is a collaborative effort.

7.13 Why might it be important to recheck instruments that were calibrated at the factory?

Movement during shipment may have affected the instruments, mistakes may have occurred at the factory, the factory's standards for calibration may be different than those used at the actual plant, errors in the ranges requested may have occurred, or changes in the process may require that changes be made to the transmitters.

7.14 Name some tools that are used to troubleshoot plant problems. Explain how they are used.

Tools used to troubleshoot plant problems include:

- *A **test gauge** – An example would be a 0–30 psig (0–207 kPC) test gauge for calibrating a control valve. Connect a regulator and a test gauge to a control valve* positioner input. Calibrate the positioner to 3–15 psig (21–103 kPC) or other specs. The test gauge is used to test a meter installed in the field (e.g., to check a pressure transmitter in a pipeline). Air pressure is applied to the line (using the five-point test) and the test gauge is read. The transmitter being checked must read within the transmitter manufacturer's acceptable tolerances to pass the test and to be accepted into plant operation.

- **A 4–20 mA current source, i.e., multimeter** – *This is used to test any device that must be driven by this current, such as a control valve or a transmitter. To use it, lift one wire off of the transmitter so a reading is removed from the control system, for example. Then connect the current source to drive the instrument and check the reading at the control system.*

See the CST tools section in Chapter 7 for more answers.

7.15   What does *mechanically complete* (MC) mean? Why might a project not wait until everything is MC to move on to the next phase of the project? Why might this involve simultaneous operations (SIMOPS)?

Mechanical completion *is a milestone in the construction phase that is detailed in the engineering, procurement, and construction (EPC) contract stating that a plant has been built per engineering specifications, all equipment has been installed (including electrical and instrumentation), and commissioning activities may commence. Depending on the size of the project, management and production demands, and the parts of the process that must be started first, MC may not be complete before commissioning starts. Because this causes simultaneous activities to occur and safety must be maintained, SIMOPS procedures must be followed.*

7.16   What is the difference between pre-commissioning and commissioning?

Pre-commissioning *normally consists of activities that must be done before chemicals can be introduced into the project locations. These activities include:*

- *Power and grounding are completed.*

- *The process automation system (PAS) is installed and connected to instrumentation.*

- *Instruments, analyzers, and gas detectors are installed and calibrated.*

- *Public address (PA)/general alarm (GA) testing is complete.*

- *Process lines and vessels are being cleaned.*

- *Equipment lubrication is completed.*

- *Refractory (firebrick) is being cured according to the manufacturer's directions.*

- *Rotating equipment and motors are checked that they turn in the proper direction.*
- *Catalyst is being charged to systems that require it.*

Commissioning *occurs after pre-commissioning activities and normally involves the following:*

- *Power is turned on (all voltage levels).*
- *Manual valves are opened and valves are stroked open/closed.*
- *Motors are run for longer periods of time.*
- *Loop checks and calibration are completed.*
- *PAS complex logic is checked (interlocks, sequences, recipes) and signed off on by qualified people.*

7.17 What are some questions a CST may ask while he or she is checking a problem with a control loop?

- *Is the control system configuration correct? This includes the tag database and graphic display.*
- *Is the signal coming from the field?*
- *Did this loop ever work correctly?*
- *Is what the operator sees on the control system actually what is happening in the field?*
- *Is the transmitter functioning properly?*
- *Are there programs, interlocks, or other connections interfering with the independent operation of this loop?*
- *When did this loop start behaving improperly?*
- *Is the final control element responding properly?*
- *Should the loop be tuned, or is there a physical problem preventing the loop from functioning properly?*
- *Is the signal going to the field?*
- *Is the loop calibrated properly?*

7.18 What tool may be used to troubleshoot an analog transmitter?

A HART communicator if the instrument or valve is HART capable, otherwise a conventional loop calibrator.

7.19 Why is vendor documentation important when troubleshooting a problem?

Vendor documentation normally contains flowcharts and/or other instructional materials for working on the equipment. It can also contain contact numbers (toll free) you can call for help. It should be the most detailed and up-to-date information on the vendor's product you are working on.

7.20 Name some things that are completed during process automation system (PAS) configuration.

- *I/O and loop database*
- *Graphics*
- *Trends*
- *Complex logic (sequences, interlocks, and complex control schemes)*
- *Historian database*

7.21 Why is it important to work with a vendor representative or equipment specialist?

Communicating with vendor representatives and specialists gives the CST the opportunity to gain expert knowledge in how a system works and how to troubleshoot it from the most knowledgeable person associated with the system.

7.22 Who are some people you can go to for assistance?

The control systems engineer, the vendor, your manager, and the operator can provide assistance.

7.23 What is the difference between a safety instrumented system (SIS) and a safety instrumented function (SIF)?

- *SIS – A control system used for (emergency) safety protection*
- *SIF – Logic (software) used by the SIS to perform the safety protection*

7.24 What plant assistance might be required during start-up?

Contractors with special expertise or skills, vendor or service representatives with information about certain pieces of equipment, and personnel from sister plants with experience in the same process might be required during start-up.

Chapter 8

8.1 Name some safety checks that must occur before the start-up of any equipment or system.

A pre-startup safety review (PSSR) has been completed and there is a start-up plan in force that has been communicated to all parties.

8.2   Why does a start-up team meet more frequently as the start-up time approaches?

Things begin to come together toward the end of construction, and more activities begin to overlap. Other groups of people become involved (e.g., operators), and pressure begins to mount as management gets ready to start production in a safe and efficient manner.

8.3   Give an example of a processing problem.

Check for tags and locks, blinds or blanks, acid flange covers, insulation; check that rupture disks and pressure relief valves are installed; check that there are self-contained breathing apparatuses (SCBA), fire extinguishers, and people trained in emergency procedures. Have a list of emergency contacts readily available.

8.4   What is meant by *troubleshooting*, and what are some of the tasks involved?

Troubleshooting is the process of identifying the source of a problem, determining the appropriate corrective action, correcting the problem, verifying the functionality of the device or system, and documenting the actions taken. This may be conducted by an individual or a team. As a CST, your forte should be troubleshooting because you have advanced training in electronics, instrumentation, and control: you are familiar with the equipment; and you have experience in solving problems encountered in the plant project from design through start-up.

8.5   Name some environmental checks done while completing a pre-startup safety review (PSSR).

- *Are containment facilities (dikes) adequate?*
- *Have arrangements been made for disposal of waste materials?*
- *Is there a chemicals inventory list?*
- *Is there a spill standard operating procedure (SOP)?*
- *How have material loading/unloading facilities been constructed?*
- *Have all waste streams been identified, quantified, analyzed, and minimized?*
- *Are all of the applicable environmental and operating permits in place?*

8.6   What things might hinder start-up and thus affect staffing?

Plant size, plant coverage (hours worked), the timetable for start-up, and the number of loops might affect staffing. Fewer people may be necessary during the very early or later stages of the start-up.

8.7 What is a *performance guarantee*?

A **performance guarantee** *is a contract made between a supplier of the project design or a piece of equipment within the project. The performance guarantee states that the item can meet certain specifications (e.g., production rate) and once this performance milestone is met or once a piece of equipment has met the performance specifications, final payment can be made to the supplier.*

8.8 Why is personal protective equipment (PPE) an important part of a PSSR? What other items, for example, process area checks, are important to review during a PSSR?

It is important to include PPE in the PSSR to ensure that personnel are equipped with safety equipment that is appropriate for the tasks they are working on, that they know how to use and take care of it, and that procedures related to PPE are understood and upheld.

Other items that are important to check during a PSSR are:

- *General safety*
- *Machinery and equipment safety*
- *Ergonomics*
- *Occupational health*
- *Process hazard analysis (PHA)*
- *Mechanical integrity*
- *Environmental protection*
- *Training*
- *Emergency response*
- *Field or process area concerns*

8.9 What may a CST do during configuration?

- *Assist with instrument connections to the process automation system (PAS)*
- *Calibrate and loop check with the CSE while he/she is performing configuration*
- *Learn how the logic works in order to understand whether hardware or software is the problem*
- *Understand how to use the PAS to determine how instruments are wired*

8.10   What are some examples of special equipment tasks performed during start-up?

Some examples of special equipment tasks include:

- *Checking that pump seal water has been connected and is functioning*
- *Verifying lubrication has been completed and motor direction is correct*
- *Removing blinds*
- *Installing safety devices*
- *Curing refractory*

8.11   What is meant by *steady-state operation*?

Steady-state operation *is when the process is running smoothly and with minimal variation in process conditions (i.e., temperature, flows and production rate, and product quality).*

8.12   What factors should be considered in connection with power in preparing for start-up?

Power must be supplied to all field instrumentation and equipment. Grounding must be properly installed and connected. Starters should be energized, rotation verified, fuses installed, wires terminated, and locks and tags removed.

8.13   Name some post-project activities that might involve a CST.

- *Working with personnel contracted to maintain the PAS*
- *Replacing instrumentation or PAS parts as needed*
- *Inventorying spare parts in the warehouse or getting these parts from the warehouse to use in the field as needed*
- *Performing maintenance tasks that emanate from the work order system*

8.14   Name some factors that might influence the order of start-up.

The order of start-up is influenced by many factors, including the schedule, the completion of installation, and the PAS configuration.

Appendix E: ISA Form 20.50

CONTROL VALVE DATA SHEET

PROJECT _____ DATA SHEET _____ OF _____
UNIT _____ SPEC _____
P.O. _____ TAG _____
ITEM _____ DWG _____
CONTRACT _____ SERVICE _____
MFR. SERIAL* _____

#			Units	Max Flow	Norm Flow	Min Flow	Shut-Off
1	Fluid				Crit Press PC		—
2		Flow Rate					
3	SERVICE CONDITIONS	Inlet Pressure					
4		Outlet Pressure					
5		Inlet Temperature					
6		Spec Wt. / Spec Grav / Mol Wt.					—
7		Viscosity / Spec Heats Ratio					—
8		Vapor Pressure P_v					—
9		Required C_v*					—
10		Travel*	%				0
11		Allowable / Predicted SPL*	dBA				—
12							

#				#		
13	LINE	Pipe Line Size and Schedule	In _____	53		Type*
14			Out _____	54		Mfr & Model*
15		Pipe Line Insulation _____		55		Size* Eff Area
16	VALVE BODY / BONNET	Type* _____		56		On / Off Modulating
17		Size* _____ ANSI Class _____		57		Spring Action Open / Close
18		Max Press/Temp _____		58	ACTUATOR	Max Allowable Pressure*
19		Mfr & Model* _____		59		Min Required Pressure*
20		Body / Bonnet Matl* _____		60		Available Air Supply Pressure:
21		Liner Material / ID* _____		61		Max Min
22		End Connection	In _____	62		Bench Range* /
23			Out _____	63		Actuator Orientation
24		Flg Face Finish _____		64		Handwheel Type
25		End Ext / Matl _____		65		Air Failure Valve Set At
26		Flow Direction*		66		
27		Type of Bonnet*		67		Input Signal
28		Lub & Iso Valve Lube		68	POSITIONER	Type*
29		Packing Material*		69		Mfr & Model*
30		Packing Type*		70		On Incr Signal Ouptut Incr / Decr*
31				71		Gauges By-Pass
32	TRIM	Type*		72		Cam Characteristic*
33		Size* Rated Travel		73		
34		Characteristic*		74	SWITCHES	Type Quantity
35		Balanced / Unbalanced*		75		Mfr & Model*
36		Rated* C_v F_L X_T		76		Contacts / Rating
37		Plug / Ball / Disk Material*		77		Actuation Points
38		Seat Material*		78		
39		Cage / Guide Material*		79	AIRSET	Mfr & Model*
40		Stem Material*		80		Set Pressure*
41				81		Filter Gauge
42				82		
43	SPECIALS / ACCESSORIES	NEC Class Group Div		83	TESTS	Hydro Pressure*
44				84		ANSI / FCI Leakage Class
45				85		
46				86		

Rev	Date	Revision	Orig	App

* Information supplied by manufacturer unless already specified

Appendix F: Acronyms

AIChE	American Institute of Chemical Engineers
AMS	asset management systems
ANSI	American National Standards Institute
API	American Petroleum Institute
ASME	American Society of Mechanical Engineers
ASTM	American Society for Testing and Materials
BMS	burner management systems
BPCS	basic process control system
C&E	cause and effect
CAD	computer-aided design
CAER	community awareness and emergency response
CAP	Certified Automation Professional
CBT	computer-based training
CCPS	Center for Chemical Process Safety
CCST	Certified Control Systems Technician
CEMS	continuous emission monitoring system
CFR	Code of Federal Regulations
CM	construction management
CO	carbon monoxide

CSE	control systems engineer
CSP or CSFP	Certified Safety Professional or Certified Safety Functional Professional
CST	control systems technician
CWHSSA	Contract Work Hours and Safety Standards Act
dB	decibel
DCS	distributed control system
dP	differential pressure
DVM	digital volt-ohm meter
EDMS	electronic document management system
EHS	Environment, Health and Safety
EPA	Environmental Protection Agency
EPC	engineering, procurement, and construction
ERP	enterprise resource planning
ERT	emergency response team
ESD	emergency shutdown
ETV	Environmental Technology Verification
EU	engineering unit
FAT	factory acceptance test
FDA	Food and Drug Administration
FEED	front-end engineering and design
FPS	fire protection system
FRC	fire-retardant clothing
FDM	field device management
GA	general arrangement
GDP	good documentation practices
GMP	good manufacturing practices
HART	highway addressable remote transducer
HAZOP	hazard and operability study
HAZWOPER	Hazardous Waste Operations and Emergency Response
HMI	human-machine interface
HR	Human Resources
I&E	instrument and electrical

IHS	information handling services
I/O	input/output
I/P	current to pneumatic
IEC	International Electrotechnical Commission
IEEE	Institute of Electrical and Electronics Engineers
ISA	International Society of Automation
ISO	International Standards Organization
ISPE	International Society for Pharmaceutical Engineering
IT	information technology
JSA	job safety analysis
kPa	kilopascal
LDAR	leak detection and repair
LLC	limited liability company
LOPA	Layer of Protection Analysis
LOTO	lockout/tagout
LSH	level switch high
mA	milliamp
MAC	main automation contractor
MCC	motor control center
MMPS	machinery monitoring and protection system
MOC	Management of Change
MSDS	Material Safety Data Sheet
NDT	nondestructive test
NFPA	National Fire Protection Association
NOx	nitrogen oxide
O_2	oxygen
OCR	optical character recognition
OLE	Object Linking and Embedding
OPC	Open Platform Communications
OSHA	Occupational Safety and Health Administration
P&ID	piping and instrumentation drawing
P/I	pneumatic to current
PAS	process automation system

PC	personal computer
PEMS	predictive emission monitoring system
PFD	process flow diagram
PHA	process hazard analysis
PID	proportional-integral-derivative
PIN	plant information network
PLC	programmable logic controller
PM	preventive maintenance
PM	project manager
PO	purchase order
PPE	personal protective equipment
PRD	pressure relief device
PSDS	Product Safety Data Sheet
psig	pounds per square inch gauge
PSM	process safety management
PSSR	pre-startup safety review
PV	process variable
QA/QC	quality assurance/quality control
QSR	quality systems regulations
RTD	resistance temperature detector
SAP	Systemanalyse und Programmentwicklung (Systems, Applications, and Products in Data Processing)
SARA	Superfund Amendments and Reauthorization Act
SAT	site acceptance test
SCBA	self-contained breathing apparatus
SCFM	standard cubic feet per minute
SDS	Safety Data Sheet
SI	system integrator
SIF	safety instrumented function
SIMOPS	simultaneous operations
SIS	safety instrumented system
SIT	site integration test
SO_2	sulfur dioxide

SOP	standard operating procedure
SP	set point
TAC	technical assistance center
THF	tetrahydrofuran
TPPS	third-party packaged system
UL	Underwriters Laboratories
VDC	voltage direct current
WO	work order

Index

Note: Page numbers followed by n indicate chapter footnotes.

A

alarms, 26, 114, 129, 143, 169, 181
annotations, 54
ANSI/ISA 95, 6, 104, 109
asset management system (AMS), 129, 180
audits, 7, 27, 173–174, 210
availability, 38, 57, 170, 174
availability and performance guarantee test, 79, 84, 125

B

backup, 104, 130, 172, 196, 199
barricades, 31
baseline tasks/work, 14–15, 206
basic process control system (BPCS), 12, 67, 79, 109, 128, 148, 170, 202, 215
batch, 105, 150, 167
benchmarks, 174
beneficial operation, 174
as-built, 3, 13, 14, 54, 73, 86, 113, 116, 175, 217
brownfield start-up, 1
bumping, 95
burner management systems, 18, 42, 126, 207
business network, 47, 73, 80, 130

C

calibration data sheet, 69, 70, 85, 116
calibration certificates, 139
cascade control scheme, 85, 116, 129
cause-and-effect (C&E) matrix, 74, 114

CEMS. *See* continuous emission monitoring system (CEMS)
Certified Automation Professional (CAP), 176–177
Certified Control System Technician (CCST), 5, 22, 98, 103, 176
Certified Functional Safety Professional (CFSP)/ Certified Safety Professional (CSP), 18
chain of command, 106, 110–111
checklists, 77, 78, 117, 123, 161, 195
Code of Federal Regulations (CFR), 20, 37, 77, 160
commissioning, 113, 138, 147–149, 156, 158, 160, 213, 220, 222
communications protocols, 130, 197
compliance, 20, 24, 60, 101, 140, 197, 202, 221
compliance audits, 20, 27
compliance documentation, 19, 36–37
computer-based training (CBT), 29, 31
configuration training, 128–130
confined space, 32, 162
construction, 3–4, 18, 49, 62, 92–93, 96, 107, 114, 122, 137–139, 141, 222
continuous emission monitoring system (CEMS), 100, 101, 131
contractors, 24, 107
Contract Work Hours and Safety Standards Act (CWHSSA), 122
control logic, 73–74, 81, 157
control systems engineer (CSE), 95–96, 127, 149, 167, 170, 215, 224
control systems technician (CST), 1–2, 6–14
converter, 73, 95, 145

237

Coriolis, 85
critical path/Gantt chart, 57–58, 93, 98, 108, 123, 126, 137, 149, 150, 156, 166, 171, 210, 213, 217, 219

D
data sheet, 12, 85, 230. *See also* specification sheet
delays, 46, 58, 74, 125, 148, 168, 171, 213
deliverables, 3, 70
design and engineering assistance, 13
detailed design, 3
distributed control system (DCS), 5, 29, 58, 109, 142, 170
documentation, 13
Document Control, 47, 87, 96, 116, 118–119, 141
document locations, 56–57
drafters, 49, 84
dry run, 157
duties, 9–10, 14–15, 23, 33, 91, 96–97, 99–100, 102, 108, 111, 123, 140, 206

E
electrical safety, 31–32
electrical wiring diagram/one-lines, 71, 72, 73, 86–87
electronic drawings, 48–49
electronics, 22, 55, 225
emergency action plan, 26
Emergency Communication Guide, 133
emergency contacts, 133–134
emergency drill, 40
emergency preparedness, 18, 26, 210
emergency response, 26
emergency shutdown (ESD) testing, 165–166, 169
emissions, 100–102, 131, 158
employee involvement, 21
employee orientation, 4, 26
employee training, 7, 23–24
energy balance, 61–62, 97, 212
engineering, procurement, and construction (EPC) firms, 92, 222
enterprise network, 79, 95, 104, 109, 125
environment, 5, 12, 18, 29, 100, 143, 147, 176, 219
environmental department, 99–102, 214
Environmental Protection Agency (EPA), 100
Environmental Technology Verification (ETV), 100, 101
errors, 17, 18, 69, 80, 107, 116, 118, 125, 130, 131, 144, 170, 211, 221
evacuation, 26, 40, 143, 158

F
Fair Labor Standards, 122
factory acceptance testing (FAT), 3
FAT certificate, 77
FAT documents, 79–80
FAT plan, 46, 79–80, 81
FAT test results, 77, 84
feedstock, 165, 168
field device management (FDM), 129, 141, 146
firewall, 73
flowchart, 52, 75, 76, 224
fiber optic, 3, 104, 128, 131, 145, 156
fieldbus, 51, 128, 131, 146
Food and Drug Administration (FDA), 20, 83, 97, 98
front end engineering and design (FEED), 3
functional specification, 58, 84, 174
functional tests, 81

G
Gantt chart/critical path, 57–58, 93, 98, 108, 123, 126, 137, 149, 150, 156, 166, 171, 210, 213, 217, 219
general arrangement drawing, 59, 63, 64, 84, 216
good automated manufacturing practice, 82
good documentation practices (GDP), 82
good engineering practice, 21
good manufacturing practices (GMPs), 20
greenfield start-up, 1, 18
grounding, 4, 140, 142, 156, 196, 219, 222, 227

H
halon, 38
hard-copy, 37, 45, 47, 86, 100
hazard communication, 8, 24, 32, 38
HAZOP (hazard and operability studies), 62, 118
hearing protection, 32, 38
HMI (human-machine interface), 68, 93, 141, 147–148, 200
housekeeping, 10–11
human-machine interface. *See* HMI
human resources, 103–104

I
incident, 9, 26, 32, 125
incident investigation, 26
industry practice, 22, 211
Information Technology (IT), 80, 96, 104–105
inspection, 3, 5, 10, 25, 77, 82–86, 110–111, 116, 126, 137, 139, 152, 180, 198
installation detail, 69–70
installation training, 128
instrument repair technician, 179–182
instrument specification/data sheet, 12, 65–66, 196, 216
intelligent instruments, 51, 129, 131, 141
International Electrotechnical Commission (IEC), 20
International Standards Organization (ISO), 20, 49, 98

J
job briefing, 28
job descriptions, 4–6, 179–182, 206
job safety analysis (JSA), 23, 34, 145
jumpering out, 42, 207

K
key performance indicators (KPI), 174

L
ladders, 6, 32, 73, 74, 213, 216
Layer of Protective Analysis (LOPA), 22
layers, 52, 211
leader, 13–14
leadership, 34
lead sheet, 49
lead tech, 6, 10, 123, 218
leak detection and repair (LDAR), 102
legend sheet, 49, 50, 54, 63
liaison, 8–9
line-breaking, 32
lines, 52
lockout, 34
lockout/tagout (LOTO), 8, 18, 32, 34–36, 41, 142
logic diagram, 73–74
logic solver, 148
loop check/log sheet, 68, 69, 85, 86, 150, 151, 217
loop check/loop checking, 6–7, 10–12, 14, 35, 46, 58, 65, 68–69, 77, 83, 85–86, 95, 99–100, 123, 126–127, 134, 138–139, 141, 143–145, 147, 149–151, 173, 206, 208, 216–220, 223
loop folder, 3, 14, 69, 77, 83–86, 139–141, 147, 149–150, 217
loop sheet/diagram, 3, 35, 67, 85, 114, 115, 131
loop tuning, 138, 166, 170–172

M
machinery monitoring and protection system (MMPS), 141–142, 158
maintenance department, 99
maintenance training, 130–131
Management of Change (MOC), 11, 12, 22, 25, 54, 97, 113, 114–118, 149, 209, 216
manufacturers' information, 70–71, 85
manufacturing, 93, 105
mass or material balance, 61–62
maximum achievable control technology, 102
mechanical completion, 92, 142, 222
mechanical integrity, 25, 221
multidisciplinary team, 26, 81, 118, 156

N
National Emission Standards for Hazardous Air Pollutants (NESHAP), 102
National Fire Protection Association (NFPA), 22
near miss, 26, 32, 103, 145, 176, 209
network topology, 73, 80
nonroutine maintenance, 8, 22, 23, 25

O
Object Linking and Embedding (OLE), 128n1
Occupational Safety and Health Administration (OSHA), 8, 20–27, 31–32, 37–38, 77, 113, 118, 130, 184–194, 210, 216, 221
Open Platform Communications (OPC), 128, 130, 131
one-lines, 71
on-stream, 157
on-the-job training (OJT), 127
operating procedures, 23
operator training simulator (OTS), 32
OSHA PSM, 8, 19, 20–27, 28, 41, 208, 216, 221

P
packaged systems, 107
partial stroke testing (PST), 130
pass/fail, 80, 81
performance guarantees, 79, 125, 159, 174, 226
performance test, 4, 174
personal protective equipment (PPE), 6, 30, 38, 187, 208, 226
piping and instrumentation drawing (P&ID), 23–25, 35, 47–52, 62–63, 86, 107, 114, 207, 210, 216
planner/scheduler, 58, 98
plant information network, 47
plot plan, 52, 63, 65
pneumatic, 34, 49, 52, 62, 67, 73, 95, 126, 145, 157, 179–180, 210, 212, 219
policy, 36, 87, 119, 122, 123
pre-commissioning, 4
predictive emission monitoring system (PEMS), 100, 102
prestart-up, 24–25, 137–152
pre-startup safety review (PSSR), 4, 24, 41, 77, 158, 160–161, 176, 208, 224, 225, 226
process automation system (PAS), 5, 18, 58, 79, 93, 116, 125–126, 137, 155, 205, 219, 224
process control narrative, 74
process engineering, 95–97
process flow diagram (PFD), 55–56, 59–61, 114–115, 207, 216
process hazard analysis (PHA), 7, 42, 39–40, 160, 207, 216
process hazard evaluation, 22
process modeling, 171
process simulation, 171
process validation, 81–82
production, 60–61, 93, 98, 105, 114, 166

PROFIBUS, 146
programmable logic controller (PLC), 5, 73, 109, 142
project, 1, 7, 9–10, 12–14, 17–19, 22–24, 26, 34, 45, 56–58, 61, 66, 69, 83, 91–93, 104, 106–108, 113, 123, 125–128, 142–144, 147–149, 155–156, 159–160, 166–169, 174, 179
project close-out, 175–176
project documents, 46, 87, 118–119
project engineering, 96–97
project execution and control, 3–4
project management/management personnel, 3–4, 98–99
project manager (PM), 92, 98–99, 116, 126, 149, 156
project review meeting, 85, 98, 137, 139, 140, 213–214
project schedule, 57, 98, 126, 137, 156, 166–168, 213
project team assignment, 94
proportional-integral-derivative (PID) control, 116, 129, 170
punch list, 77, 81, 138, 139–140
Purchasing department, 102–103

Q
quality/inspection, 10–13, 77
quality management systems, 97
quality systems regulations (QSRs), 20

R
ramp up, 165, 169
read-only, 56
record keeping, 158
regulatory safety training, 27, 28, 31–32
respirator, 40
revisions, 54
Right-to-Know, 24, 37
risk reduction, 41, 148
routine maintenance, 1, 130, 179

S
Safety Data Sheet (SDS), 37–38, 184–194
safety equipment, 38–39, 158, 170, 226
safety inspector, 7–8
safety instrumented function (SIF), 41, 148n7, 224
safety instrumented system (SIS), 5, 41, 67, 128, 157, 224
safety meetings, 28–31
safety stand-down, 28
safety training, 31–34
SAP (Systems, Applications, and Products in Data Processing), 6
scale, 52
schedule, 14–15, 57–58, 71, 86–87, 96, 98, 108, 124–127, 137, 156, 166–168, 171, 203, 213

security, 56, 73, 104, 128, 196, 199
shelter-in-place, 40, 158
simultaneous operations (SIMOPS), 4
SIS. *See* safety instrumented system
site acceptance test (SAT) plan, 3, 77, 81, 195
site integration test (SIT) plan, 77, 80, 195
smart instrumentation, 129, 142
sniffers, 35, 38
specification sheet, 12, 63, 65–66, 71, 85, 217, 220. *See also* data sheet
staffing, 121–127, 166–167, 218, 225
staging area, 4, 79
standard operating procedure (SOP), 29, 30, 56, 75, 77, 114, 207, 211, 220
start-up, 1, 77, 84, 155–160, 165–172
steady-state, 166, 169–171, 174–175, 227
stroking, 95
Superfund Amendments and Reauthorization Act (SARA), 37, 313
symbols, 54
system diagnostics, 130, 199
system integrator, 74, 79, 108–109, 129, 146, 167

T
tagging, 198, 220
tagout, 35
tagname, 85
technical lead, 106, 110, 123
temporary changes, 52, 114, 211
test gauge, 145, 221–222
testing, 3, 4, 29, 46, 80, 83, 130, 137–138, 148, 157, 165–166, 169–170, 172, 174, 195
testing documentation, 77–79
test methodology, 80
third-party packaged system (TPPS), 67, 71, 73, 79, 95, 106, 107–109, 128, 130, 137, 142, 146
toolbox, 28
toxic chemical release inventory, 37
TPPS. *See* third-party packaged system
trade secrets, 27
training, 7, 23–24, 27, 28, 29, 31–34, 127–133
troubleshooting, 95, 122–123, 127–128, 138, 143, 169, 177, 180, 225
tuning, 166, 170–172

V
validation, 81–83, 97

W
welding, 99, 139, 168, 197
wet run, 157
wiring, 11, 22, 49, 71–73, 86, 99, 107, 116, 119, 127–128, 132, 140, 145, 180, 196–198

Printed and bound by CPI Group (UK) Ltd, Croydon, CR0 4YY
22/04/2026
14866382-0004